A Compendium of Solid State Theory

Ladislaus Alexander Bányai

A Compendium of Solid State Theory

Second Edition

Springer

Ladislaus Alexander Bányai
Institut für Theoretische Physik
Goethe Universität
Frankfurt am Main
Germany

ISBN 978-3-030-37361-0 ISBN 978-3-030-37359-7 (eBook)
https://doi.org/10.1007/978-3-030-37359-7

This Springer imprint is published by the registered company Springer Nature Switzerland AG.
The registered company address is: Gewerbestrasse 11, 6330 Cham, Switzerland

Preface to the Second Edition

Rereading my own text some time after the publication, I felt the need to add certain supplementary material. Although it was difficult to do it within a few pages, I introduced a short description of the many-body adiabatic perturbation theory including Feynman diagrams. This is not intended to teach the respective techniques largely described in many textbooks, but at least to get a vague idea about them. It may serve also as a memo refreshing critically the basic ideas for those who already are acquainted with it. The chapter about transport theory got more important extensions. The solvable model of an electron in a d.c. field interacting with both optical and acoustical phonons has now been discussed in more detail, since it is very important for understanding irreversibility and dissipation. A subsection about the nonmechanical kinetic coefficients and another one about the derivation of the Seebeck coefficient in hopping transport typical for amorphous semiconductors were added. I also felt it necessary in the "Optical Properties" chapter to give an example about the proper use of the linear response in the presence of Coulomb interactions, illustrated by the derivation of the Nyquist theorem. The chapter on phase transitions was largely extended. It includes now a description of the Bose condensation in real time within the frame of a rate equation, as well as the excitation spectrum of repulsive bosons within Bogoliubov's s.c. model at zero temperature. The extension to its time-dependent version leads after a next simplifying approximation to the Gross–Pitaevskii equation for the condensate. The discussion of the microscopic model of superconductivity was supplemented with that of the Bogoliubov–de Gennes equation. The book ends now with two new chapters giving a broader view of the electrodynamics of the particles in the solid state. One of them is an extension of the basic solid-state Hamiltonian now including current–current interactions of order $1/c^2$ starting from the classical electrodynamics of point-like particles. The second one is a field-theoretical Lagrangian formulation of non-relativistic QED, which is necessary to understand both classical and quantum mechanical electrodynamics. Its $1/c^2$ approximation on states without photons justifies the previous approach. Of course, some improvements in the text have been made, as well as some new figures were introduced, wherever I felt it necessary.

I kept the original idea of this book to restrict the discussion to self-consistent topics, which may be clearly presented to a graduate student. In this sense, I omitted several important new developments.

Oberursel, Germany
October 2019

Ladislaus Alexander Bányai

Preface to the First Edition

This compendium emerged from my lecture notes at the Physics Department of the Johann Wolfgang Goethe University in Frankfurt am Main till 2004 and does not include recent progresses in the field. It is less than a textbook, but rather more than a German "Skript." It does not include a bibliography or comparison with experiments. Mathematical proofs are often only sketched. Nevertheless, it may be useful to graduate students as a concise presentation of the basics of solid-state theory.

Oberursel, Germany
August 2018

Ladislaus Alexander Bányai

Contents

The original version of this book was revised: author's affiliations has been updated. The correction to this book is available at https://doi.org/10.1007/978-3-030-37359-7_13

Chapter 1
Introduction

A short presentation is given about how we conceive theoretically the solid state of matter, namely about its constituents and their interactions. The starting point is of course an oversimplified one, which afterwards one tries to improve and even redefine it. In the opposite sense, one tries later to simplify it in order to allow for theoretical calculations. In spite of all these problems, one gets predictions that often may be successfully confronted with experiments. The different chapters of this compendium are, however, intended only to introduce the reader into the basic concepts of solid-state theory, without the usual detours about specific materials and without comparison with experiments.

Under solid state we understand a stable macroscopic cluster of atoms. The stability of this system relies on the interaction between its constituents. We know today that molecules and atoms are built up of protons, neutrons, and electrons. According to the modern fundamental concepts of matter at their turn these particles are made up, however, of some other more elementary ones we do not need to list here. The simplest starting point of solid-state theory and still the only useful one is that we have a quantum mechanical system of ions and valence electrons with Coulomb interactions between these particles. This non-relativistic picture of a system of charged particles, however, is valid only up to effects of order $1/c^2$ (c-being the light velocity in vacuum). Actually, the charged particles are themselves sources of an electromagnetic field, which on its turn is quantized (photons). Fields and particles have to be treated consequently together as a single system. Nevertheless, most of the properties of solid state are successfully described by a non-relativistic quantum mechanical Hamiltonian

L. A. Bányai, *A Compendium of Solid State Theory*,
https://doi.org/10.1007/978-3-030-37359-7_1

$$H = H_e + H_i + H_{ee} + H_{ii} + H_{ei} \ .$$

The electron and ion Hamiltonians H_e and H_i include also the interaction with external (given) electromagnetic fields, while the interaction parts (electron–electron H_{ee}, ion–ion H_{ii} and electron–ion H_{ei}) are understood as pure Coulomb ones.

Of course, one cannot ignore the spin as a supplementary degree of motion. Sometimes, it is nevertheless compelling to include other effects of higher order in the inverse light velocity $1/c$. Even more, in the treatment of magnetic properties one has to redefine the model by including also the spin magnetic moment of localized electrons of atomic cores. Furthermore, it seems plausible that for the understanding of magnetic phenomena one reaches the limitations of today's solid-state theory by ignoring the magnetic field produced by the currents in the solid.

An essential role in the mathematical treatment of this system plays the thermodynamic limit, i.e., letting the number of particles and the volume of the system tend to infinity, while keeping the density of particles constant. The very existence of this limit for interacting quantum mechanical particles is not at all obvious, but necessary for the stability of matter.

To treat such a still extremely complicated system it is necessary to make further simplifications. Since most solids are crystals, one admits that only the valence electrons are allowed to move over the whole crystal, while the ions at most oscillate around their equilibrium positions in the given lattice. In a first step, one starts from the model of rigid ions (their mass is thousands of times heavier than the electron mass!) in a given periodical lattice and considers the motion of the electrons in a periodic field. Actually the characteristics of the lattice should be also determined by the above Hamiltonian, but one springs over this step. The task is still too complicated, and one treats not the many electron system, but just one electron in the field of ions and the self-consistent field of the other electrons. In a first step, one considers this potential as a given one. This oversimplified picture is already able to describe qualitatively the fundamental properties of solids. This is the *one-electron theory* of solid state, which will be described in Chap. 2 of this compendium. We make here also a first step toward the many-body treatment by considering many, but non-interacting electrons within the second quantization scheme in metals and semiconductors.

In Chap. 3 we consider electron–electron interactions and many-body approximation schemes. Lattice oscillations and their interaction with the electron are discussed in Chap. 4. An important part of traditional solid-state theory concerns transport and optical properties. These will be presented, respectively, in Chaps. 5 and 6. We describe there also some new aspects related to the interaction with strong ultra-short laser pulses. Phase transformations, one of the most fascinating properties of solids we discuss in some important examples in Chap. 7. As a single modern subject we discuss in Chap. 8 some exotic properties of two-dimensional semiconductor structures, without touching the intricate theories of the quantum Hall effect. My policy was through all this compendium to describe only the most transparent theories, without claiming completeness.

As a complement to this introduction we round up in Chap. 9 the review of solid-state theory with a glimpse at a further possible extension including $1/c^2$ magnetic current–current interaction terms in the basic Hamiltonian. A deeper understanding of the many-body theory of solid state, including also its previous extension requires, however, an insight into the non-relativistic quantum electrodynamics (QED). This is made accessible within a field theoretical Lagrangian approach in Chap. 10.

The compendium ends with two appendices. In Chap. 11 we give an overview of the concepts of theoretical physics we use. It is only a reminder and it is supposed that the reader is familiar with all of them. Further, in Chap. 12 some homeworks are proposed for the interested reader.

Chapter 2
Non-Interacting Electrons

Most properties of a crystal may be interpreted as the quantum mechanical motion of a single electron in a given potential. After the simplest cases of motion in homogeneous electric and magnetic fields with emphasis on the thermodynamic limit, an extended treatment of the motion in a periodical potential is given. The Bloch oscillations observed in periodical semiconductor layers are also included. The ground state occupation of the one-electron states leads to the understanding of the different classes of materials, as metals, insulators, and semiconductors. The properties of the latter are strongly influenced by the presence of impurities. The controlled presence of donors and acceptors determines the properties of semiconductor contacts discussed on the example of a p–n contact.

2.1 Free Electrons

In the frame of quantum mechanics, the stationary state (wave function) of a free electron in the whole space is described by a plane wave

$$\psi(\mathbf{x}) = e^{\imath \mathbf{kx}}$$

having a continuous energy spectrum (kinetic energy) depending only on $k = |\mathbf{k}|$

$$\epsilon(k) = \frac{\hbar^2}{2m}k^2 \qquad (0 < k < \infty).$$

L. A. Bányai, *A Compendium of Solid State Theory*,
https://doi.org/10.1007/978-3-030-37359-7_2

This state is not normalized

$$\int d\mathbf{x}|\psi(\mathbf{x})|^2 = \infty$$

i.e., it is not a true eigenfunction.
Since we want to deal with true eigenfunctions and to consider systems with a definite volume (for simplicity a cube of size L^3) we may look for eigenfunctions in the product form $\psi(\mathbf{x}) = \phi_1(x_1)\phi_2(x_2)\phi_3(x_3)$ with two possible choices of boundary conditions in each dimension.
We may impose on the wave function $\phi(x)$ either to

(a) vanish at the boundary (Dirichlet): $\phi(\pm\frac{L}{2}) = 0$, then one gets symmetrical and anti-symmetrical eigenfuntions

$$\phi_k(x)^s = \sqrt{\frac{2}{L}}\cos(kx) \qquad \left(k = (2n+1)\frac{\pi}{L}; \quad n = 0, 1, 2, \cdots\right)$$

$$\phi_k(x)^a = \sqrt{\frac{2}{L}}\sin(kx) \qquad \left(k = n\frac{2\pi}{L}; \quad n = 1, 2, 3, \cdots\right)$$

or

(b) to be periodical: $\phi(x + L) = \phi(x)$, then the eigenfunctions are

$$\phi_k(x) = \frac{1}{\sqrt{L}}e^{\imath kx} \qquad \left(k = n\frac{2\pi}{L}; \qquad n = \pm 1, \pm 2, \cdots\right).$$

The energy in both cases is given by $\epsilon_k = \frac{\hbar^2}{2m}k^2$, however, with the above given discrete eigenvalues. The wave functions $\phi_k(x)$ are normalized in the interval $[-\frac{L}{2}, \frac{L}{2}]$.

It is convenient to work with variant (b), since these eigenfunctions are also eigenfunctions of the momentum operator

$$-\imath\hbar\frac{\partial}{\partial x}\phi_k(x) = \hbar k\phi_k(x).$$

One has orthonormality (with the “bra” and “ket” symbols of Dirac)

$$\begin{aligned} < k|k' > &\equiv \int_{\frac{-L}{2}}^{\frac{L}{2}} dx\phi_k(x)^*\phi_{k'}(x) \\ &= \frac{1}{L}\int_{\frac{-L}{2}}^{\frac{L}{2}} dx e^{(k'-k)x} = \int_{\frac{-1}{2}}^{\frac{1}{2}} dy e^{2\pi(n'-n)y} = \delta_{k,k'} \end{aligned}$$

as well as the completeness of Fourier series.

The electrons are fermions with spin $\frac{1}{2}$. The spin projection on the arbitrary chosen z-axis σ has two eigenstates with eigenvalues $\pm\frac{1}{2}$ and therefore the electron states ("kets") we have to consider are

$$|\mathbf{k}, \sigma>,$$

with the quantum numbers $\mathbf{k} = (k_1, k_2, k_3), \sigma = \pm\frac{1}{2}$. However, the energy of the free electron is spin-independent.

If one wants to consider states with many electrons, then according to the Pauli principle for fermions one must build up anti-symmetrized products of one-electron wave functions. It is, however, advantageous to use the formalism of occupation numbers characterizing a many-electron state by the occupation numbers of the one-electron states $\mathbf{k}, \sigma$. Due to the same Pauli principle these occupation numbers $n_{\mathbf{k},\sigma}$ may take only the values 0, and 1. This follows automatically from the anti-symmetry of the wave function in the configuration space. The total energy and total momentum of such a many-electron state are

$$E = \sum_{\mathbf{k},\sigma} n_{\mathbf{k},\sigma}\epsilon_{\mathbf{k},\sigma}; \qquad \mathbf{P} = \sum_{\mathbf{k},\sigma} n_{\mathbf{k},\sigma}\hbar\mathbf{k}$$

with the occupation numbers and one-electron energies

$$n_{\mathbf{k},\sigma} = 0, 1; \qquad \epsilon_{\mathbf{k},\sigma} = \frac{\hbar^2}{2m}k^2,$$

while the total number of electrons is

$$N = \sum_{\mathbf{k},\sigma} n_{\mathbf{k},\sigma}.$$

If one wants to study bulk properties of macroscopic matter (independent of the surface!), one has to perform the *thermodynamic limit* that means in this case

$$L \to \infty, \qquad N \to \infty,$$

while keeping constant the average density of electrons

$$\langle n \rangle = \frac{\langle N \rangle}{L^3}.$$

By this infinite volume limit procedure, the discrete sums go over into Riemann integrals. With $\Delta k_i = \frac{2\pi}{L}$ one has

$$\sum_{\mathbf{k}} \Delta\mathbf{k} \to \int d\mathbf{k},$$

or otherwise stated

$$\sum_{\mathbf{k}} \to \frac{L^3}{(2\pi)^3} \int d\mathbf{k}.$$

Since the one-electron energy is monotonously increasing with $k \equiv |\mathbf{k}|$, the ground state of a many-electron system is obviously the one with the one-electron states completely occupied up to a certain wave vector k_F

$$n_{\mathbf{k},\sigma} = \theta(k_F - |\mathbf{k}|).$$

The energy of an electron with this wave vector $\epsilon_F \equiv \frac{\hbar^2}{2m}k_F^2$ is called the Fermi energy.

In thermodynamic equilibrium at a given temperature and chemical potential, described by the macro-canonical distribution the probability of a many-electron state specified by the set of occupation numbers $\nu \equiv n_{\mathbf{k}_1,\sigma_1}, n_{\mathbf{k}_2,\sigma_2}, \ldots$ is given by

$$\mathscr{P}_\nu = \frac{e^{-\beta(E_\nu - \mu N_\nu)}}{\sum_{\nu'} e^{-\beta(E_{\nu'} - \mu N_{\nu'})}},$$

where $\beta = \frac{1}{k_B T}$, k_B is the Boltzmann constant, T is the absolute temperature, and μ is the chemical potential. The average occupation number of an one-electron state $\mathbf{k}, \sigma$ is then

$$\begin{aligned}\langle n_{\mathbf{k},\sigma}\rangle &= \sum_\nu \mathscr{P}_\nu n_{\mathbf{k},\sigma} \\ &= \frac{\sum_{n=0,1} n e^{-\beta(\frac{\hbar^2}{2m}k^2 - \mu)n}}{\sum_{n=0,1} e^{-\beta(\frac{\hbar^2}{2m}k^2 - \mu)n}} \\ &= \frac{1}{1 + e^{\beta(\frac{\hbar^2}{2m}k^2 - \mu)}} \equiv f(\frac{\hbar^2}{2m}k^2),\end{aligned}$$

where

$$f(\epsilon) = \frac{1}{1 + e^{\beta(\epsilon - \mu)}}$$

is the Fermi function giving the average number of electrons in a state with energy ϵ. The total average number of particles is

$$\sum_{\sigma=\pm\frac{1}{2}} \sum_{\mathbf{k}} f(\frac{\hbar^2}{2m}k^2) = \langle N\rangle$$

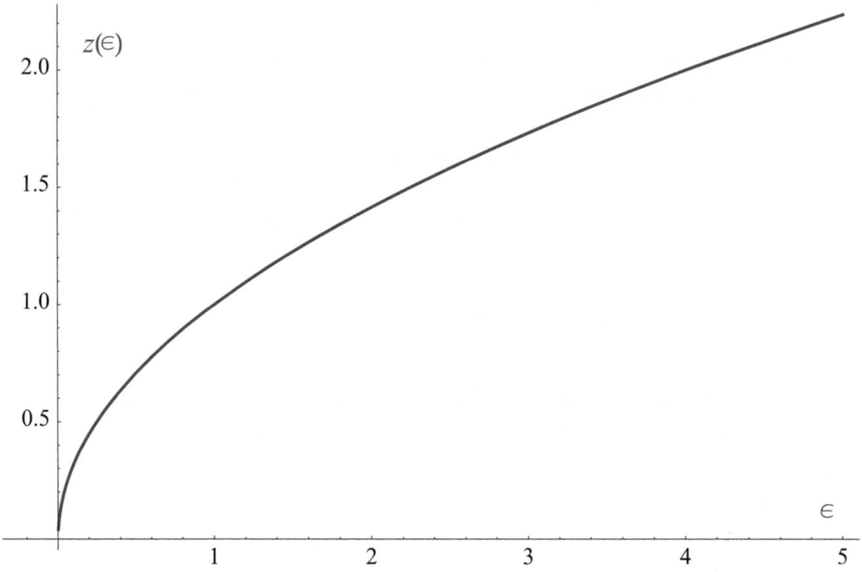

Fig. 2.1 Free one-electron state density $z(\epsilon)$

and in the thermodynamic limit (after dividing by the volume L^3 of the system)

$$\frac{2}{(2\pi)^3}\int d\mathbf{k} f(\frac{\hbar^2}{2m}k^2) = \langle n\rangle .$$

This equation determines the chemical potential μ at a given temperature T and average density $\langle n\rangle$. By introducing the one-electron state density

$$z(\epsilon) \equiv \lim_{L\to\infty}\frac{1}{L^3}\sum_{\mathbf{k},\sigma}\delta(\epsilon - \frac{\hbar^2}{2m}k^2) = \frac{2}{(2\pi)^3}\int d\mathbf{k}\delta(\epsilon - \frac{\hbar^2}{2m}k^2) = \frac{1}{2\pi^2\hbar^3}(2m)^{\frac{3}{2}}\epsilon^{\frac{1}{2}}$$

(shown in Fig. 2.1) it may be rewritten as

$$\int d\epsilon z(\epsilon) f(\epsilon) = \langle n\rangle .$$

The average electron energy is

$$\langle\epsilon\rangle = \frac{1}{\langle n\rangle}\int d\epsilon z(\epsilon) f(\epsilon)\epsilon \; .$$

One distinguishes two regimes:

(i) *degenerate*, with a big positive chemical potential $\beta\mu \gg 1$, which in the extreme case at $T = 0$ gives rise to a step-like Fermi function

$$f(\epsilon)|_{T=0} = \theta(\epsilon_F - \epsilon),$$

the Fermi energy ϵ_F being defined by $\mu|_{T=0K} = \epsilon_F > 0$ and the average one-electron energy is

$$\langle\epsilon\rangle|_{T=0} = \frac{3}{5}\epsilon_F.$$

(ii) *non-degenerate*, with $\beta\mu \leqslant 1$, which in the extreme case of a negative chemical potential with $\beta\mu \ll -1$ leads to the Boltzmann distribution

$$f(\epsilon) \approx e^{\frac{\mu-\epsilon}{k_B T}}$$

and the average one-electron energy

$$\langle\epsilon\rangle \approx \frac{3k_B T}{2}.$$

An illustration of the shape of degenerate and non-degenerate Fermi functions is given in Fig. 2.2.

2.2 Electron in Electric and Magnetic Fields

The non-relativistic Hamiltonian of an electron in the presence of given (external) electromagnetic fields is

$$H = \frac{1}{2m}\left(\mathbf{p} + \frac{e}{c}\mathbf{A}\right)^2 - eV.$$

The scalar and vector potentials $V,\ \mathbf{A}$ are related to the electric and magnetic fields by

$$\mathbf{E} = -\nabla V - \frac{1}{c}\dot{\mathbf{A}}; \qquad \mathbf{B} = \nabla \times \mathbf{A}.$$

It is important to remark that the first term in the Hamiltonian is actually the usual kinetic energy since $\dot{\mathbf{r}} = \frac{1}{m}\left(\mathbf{p} + \frac{e}{c}\mathbf{A}\right)$.

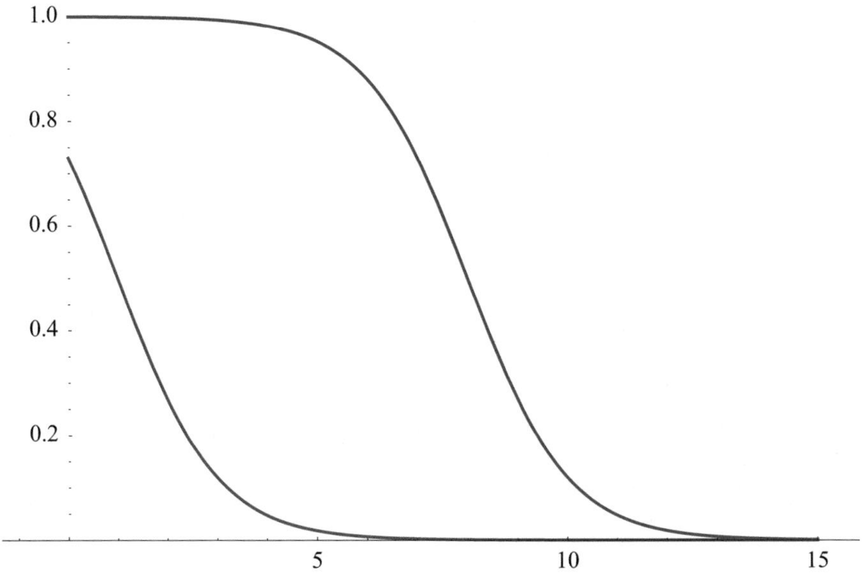

Fig. 2.2 Degenerate with $\beta\mu = 8$ (magenta) and non-degenerate with $\beta\mu = 1$ (blue) Fermi functions

2.2.1 Homogeneous, Constant Electric Field

A homogeneous, constant electric field may be described by different choices of the potentials, for example,

$$V(\mathbf{r}) = -\mathbf{rE} \qquad \textit{and} \qquad \mathbf{A} = 0$$

or

$$V(\mathbf{r}) = 0 \qquad \textit{and} \qquad \mathbf{A} = -c\mathbf{E}t.$$

In the second choice one gets a time-dependent Hamiltonian. Due to gauge invariance, the physical results are independent of the choice, but it is rather convenient to work in the first gauge with the time-independent Hamiltonian

$$H = -\frac{\hbar^2}{2m}\nabla^2 + e\mathbf{rE}.$$

Then, with $\mathbf{E} = (E, 0, 0)$ only the motion along the x-axis is affected by the electric field while the transverse motion is still described by plane waves. The solutions of the stationary Schrödinger equation with the energy $-\infty < \epsilon < \infty$

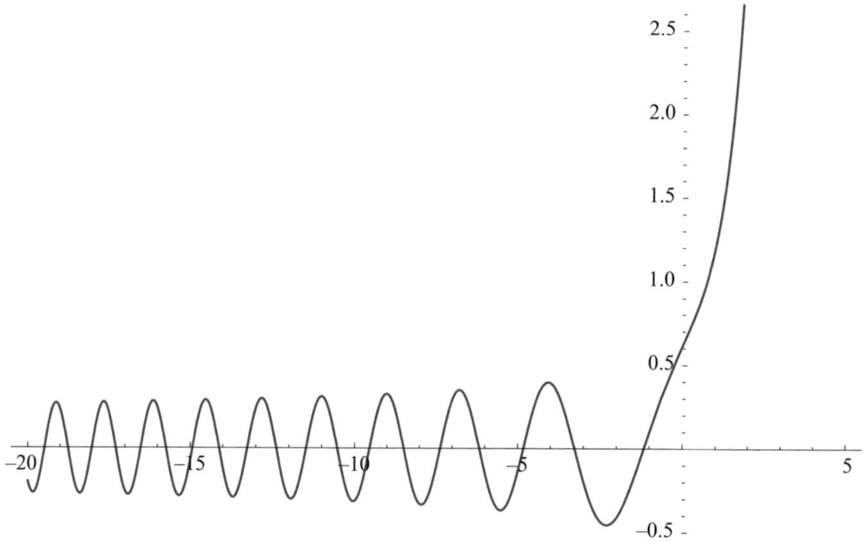

Fig. 2.3 Airy function

$$\left(-\frac{\hbar^2}{2m}\frac{d^2}{dx^2}+eEx-\epsilon\right)\phi_\epsilon(x)=0$$

are the Airy functions (seen in Fig. 2.3)

$$\phi_\epsilon(x)=\frac{1}{\sqrt{\pi}}\int_0^\infty dq\cos\left(\frac{1}{3}q^3+q\xi\right);\quad \xi\equiv\left(x-\frac{\epsilon}{eE}\right)\left(\frac{2mE}{\hbar^2}\right)^{\frac{1}{3}}.$$

Although in this gauge we found stationary states, it makes no sense to look for equilibrium since the energy spectrum is unbounded from below and these states are not eigenfunctions in the usual sense since they are not normalized

$$\int_{-\infty}^{\infty}dx|\phi_\epsilon(x)|^2=\infty$$

and the energy spectrum is continuous. This is related to the fact that already a classical electron is accelerated in a homogeneous, constant electric field. This aspect we may recover also in the quantum mechanical theory if we write the above Hamiltonian in the momentum representation:

$$H(\mathbf{p})=\frac{\mathbf{p}^2}{2m}-\imath e\hbar\mathbf{E}\frac{\partial}{\partial\mathbf{p}}.$$

Then the Heisenberg equation of motion for the momentum

$$\dot{\mathbf{p}} = e\mathbf{E}$$

follows.

2.2.2 Homogeneous, Constant Magnetic Field

Let us choose the homogeneous, constant magnetic field along the z-axis

$$\mathbf{B} = (0, 0, B)$$

and the vector potential in the Landau gauge $\mathbf{A} = (0, Bx, 0)$. Then the Hamiltonian of the electron looks as

$$H = -\frac{\hbar^2}{2m}\frac{\partial^2}{\partial x^2} + \frac{1}{2m}\left(-\imath\hbar\frac{\partial}{\partial y} - \frac{e}{c}Bx\right)^2 - \frac{\hbar^2}{2m}\frac{\partial^2}{\partial z^2}.$$

One looks for the stationary solution of the Schrödinger equation in the form

$$\psi_{k_y,k_z}(\mathbf{r}) = \Phi(x)\frac{e^{\imath(k_z z + k_y y)}}{\sqrt{L_z L_y}}.$$

Then $\Phi(x)$ has to satisfy the equation

$$\left[-\frac{\hbar^2}{2m}\frac{\partial^2}{\partial x^2} + \frac{1}{2}m\omega_c^2(x - X)^2 - \epsilon + \frac{\hbar^2 k_z^2}{2m}\right]\Phi(x) = 0.$$

As usual, one chooses periodical boundary conditions along the y- and z-axes. Therefore, $k_z = \frac{2\pi n_z}{L_z}$, $k_y = \frac{2\pi n_y}{L_y}$ with integer n_x, n_y. Here

$$\omega_c = \frac{|e|B}{mc}$$

is the frequency of the cyclotron oscillation,

$$X = sign(e)\ell_B^2 k_y$$

is the x coordinate of the center of the cyclotron motion, while

$$\ell_B = \sqrt{\frac{\hbar c}{|e|B}}$$

is the "magnetic length." As it may be seen, the problem is not completely separable, in the sense that the motion along the x-axis depends also on the quantum numbers of the motion along the other axes. The normalized (on the whole x-axis!) eigenfunctions of this Schrödinger equation are given by

$$\Phi_{n,X}(x) = \frac{1}{\sqrt{\ell_B}} e^{-x^2/2\ell_B^2} H_n((x - X)/\ell_B)$$

where $H_n(x)$ are the Hermite polynomials and the corresponding eigenenergies are quantized (oscillator values), but degenerate with respect to X

$$\epsilon_{n,X,k_z,} = \frac{\hbar^2 k_z^2}{2m} + \hbar\omega_c(n + \frac{1}{2}) \qquad (n = 0, 1, 2, \ldots).$$

The oscillator ground state eigenfunction $\Phi_{0,0}(x)$ with ℓ_B taken as unit length is shown in Fig. 2.4. In the x-direction the boundary is not fixed. One might, however, restrict the coordinate X to a certain domain $-L_x/2 \leq X \leq L_x/2$ in order to get approximate boundaries in the sense that at distances much bigger than the magnetic length far away the wave function is very small. Typical for these Landau functions is that the energy does not depend either on the quantum number X or on k_x. However, if one imposes Dirichlet (vanishing) boundary conditions at $\pm L_x/2$, the first degeneracy is lifted.

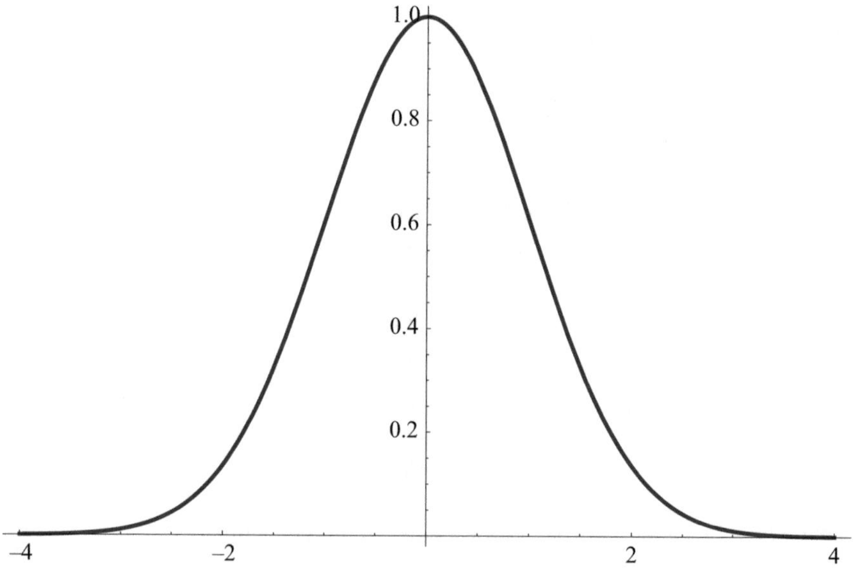

Fig. 2.4 The ground state Landau function $\Phi_{0,0}(x)$

Another convenient way to impose a "confinement" along the x axis is to introduce an oscillator potential barrier

$$V(x) = \frac{m}{2}\omega_0^2 x^2.$$

The solution is again the usual Landau function, however, the cyclotron frequency gets replaced by

$$\tilde{\omega} = \sqrt{\omega_c^2 + \omega_0^2}$$

and the cyclotron center coordinate X gets replaced by

$$\tilde{X} = \frac{\omega_c^2}{\tilde{\omega}^2} X.$$

In this way, the degeneracy with respect to X is lifted and the eigenenergies are given by

$$\epsilon_{n,\tilde{X},k_z} = \frac{\hbar^2 k_z^2}{2m} + \hbar\tilde{\omega}(n + \frac{1}{2}) + \frac{m\omega_0^2}{2}\tilde{X}^2 \qquad (n = 0, 1, 2, \ldots).$$

Nevertheless, whenever effects at the boundaries do not play an essential role, it is mostly convenient to work with the Landau functions.

In the presence of the magnetic field it is also very important to take into account that one has an additional spin-dependent energy $-2\sigma\mu_B B$, where $\mu_B = \frac{e\hbar}{2mc}$ is the Bohr magneton and the projection of the spin on the z axis may take two values $\sigma = \pm\frac{1}{2}$.

2.2.2.1 Magnetization

Since in the case of the magnetic field by restricting the cyclotron center as we mentioned before, we may construct normalized eigenfunctions with discrete spectrum bounded from below and we may look for equilibrium properties. For this sake let us consider the one-electron state density, where we took into account that the spin of the electron along the magnetic field may take two possible values $\sigma = \pm\frac{1}{2}$:

$$z(E) = \frac{1}{L_x L_y L_z} \sum_{k_z,k_y,n,,\sigma} \delta(E - \frac{\hbar^2 k_z^2}{2m} - \hbar\omega_c(n + \frac{1}{2}) + s_z 2\mu_B B).$$

By performing the thermodynamic limit we have to take into account that by limiting the range of the center of the cyclotron motion X implicitly $|k_y| \leq \frac{L_x}{2\ell_B^2}$.

Then we get in the thermodynamic limit

$$z(E) = \frac{1}{(2\pi)^2 \ell_B^2} \sqrt{\frac{2m}{\hbar^2}} \sum_{n,\sigma}^{\infty} \frac{\theta(E - \hbar\omega_c(n+\frac{1}{2}) + \sigma 2\mu_B B)}{\sqrt{E - \hbar\omega_c(n+\frac{1}{2}) + \sigma 2\mu_B B}}.$$

Obviously, one has a singular behavior at certain discrete periodical values of the energy E (see Fig. 2.5, where the unit of energy was taken to be $\hbar\omega_c$ and a red line shows the state density in the absence of the magnetic field). For sake of simplicity, the spin is ignored in the figure. Anyway, at very low temperatures all the spins are oriented along the magnetic field. This gives rise to typical oscillations of the equilibrium magnetization as function of $\frac{1}{B}$ (de Haas van Alfven effect).

Let us consider now a many-electron system in equilibrium. The equilibrium magnetization is defined as the derivative of the free energy F with respect to the magnetic field

$$M = -\frac{1}{L_x L_y L_z} \frac{\partial F}{\partial B}.$$

The free energy is related on its turn to the thermodynamic potential $\mathscr{F}$ and the average number of particles $\langle N \rangle$ by

$$F = \mathscr{F} + \mu \langle N \rangle.$$

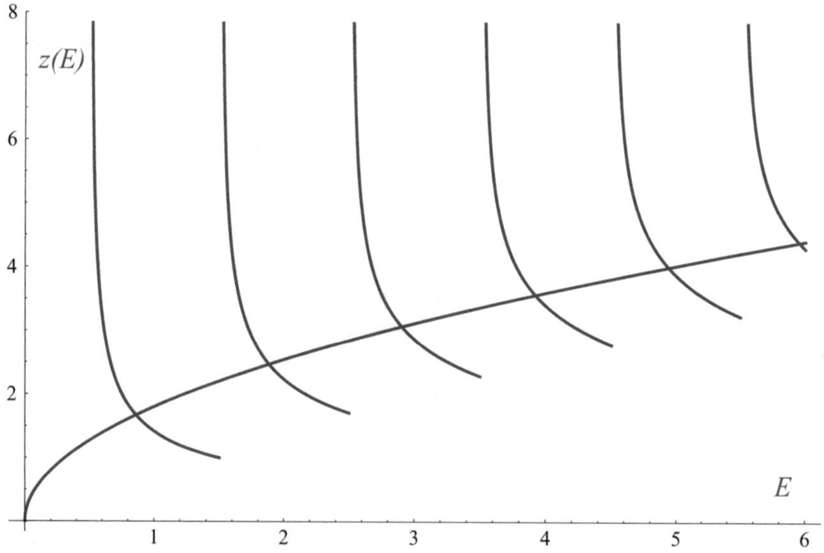

Fig. 2.5 State density $z(E)$ in magnetic field

In the macro-canonical equilibrium, we have

$$\mathscr{F} = -k_B T \ln \left\{ Sp\left(e^{-\beta(H-\mu N)} \right) \right\}.$$

With the short-hand notation i for the one-electron Landau state quantum numbers and $n_i = 0, 1$ for this occupation number we get

$$\begin{aligned}
\mathscr{F} &= -k_B T \ln \left\{ \sum_{n_1, n_2, \ldots} e^{-\beta \sum_i (\epsilon_i - \mu) n_i} \right\} \\
&= -k_B T \ln \left\{ \Pi_i \left[1 + e^{-\beta(\epsilon_i - \mu)} \right] \right\} \\
&= -k_B T \sum_i \ln \left[1 + e^{-\beta(\epsilon_i - \mu)} \right] \\
&= -k_B T L_x L_y L_z \int d\epsilon z(\epsilon) \ln \left[1 + e^{-\beta(\epsilon - \mu)} \right].
\end{aligned}$$

The average electron density $\langle n \rangle$ on its turn is defined by the one-electron state density and the average occupation number of the states given by the Fermi function

$$f(\epsilon) \equiv \frac{1}{1 + e^{\beta(\epsilon - \mu)}},$$

$$\langle n \rangle = \int d\epsilon z(\epsilon) f(\epsilon)$$

and

$$F = \int d\epsilon z(\epsilon) \left(-k_B T \ln \left[1 + e^{-\beta(\epsilon - \mu)} \right] + \mu \frac{1}{1 + e^{\beta(\epsilon - \mu)}} \right).$$

Thus, the magnetic field dependence of the one-particle state density (at a fixed chemical potential μ) determines the magnetization.

2.2.3 *Motion in a One-Dimensional Potential Well*

Let us consider here a simple one-dimensional problem of quantum mechanical motion in a static potential well. We have discussed already, how the energy spectrum of the free propagating particle can be conveniently discretized by Dirichlet boundary conditions at $x = \pm\frac{L}{2}$. Actually, this corresponds to the motion in an infinitely high potential well that does not allow the penetration inside the potential barrier. The purpose was to recuperate later the thermodynamic limit at

$L \to \infty$ while keeping the average particle density constant. Now we shall consider here a potential well of finite height U_0 and in order to avoid confusions with the previous problem we denote here the finite width of the well with a.

The stationary Schrödinger equation we consider is

$$\left\{-\frac{\hbar^2}{2m}\frac{d^2}{dx^2} - eV(x)\right\}\psi(x) = E\psi(x),$$

with the potential

$$-eV(x) = \begin{cases} 0 & for \quad |x| \leq \frac{a}{2} \\ U_0 > 0 & for \quad |x| > \frac{a}{2} \end{cases}$$

shown in Fig. 2.6.

The solutions inside the well are trigonometric functions $\sin(kx)$ and $\cos(kx)$ with $k^2 = \frac{2m}{\hbar^2}$E, while outside the well, choosing vanishing conditions at $x \to \pm\infty$ these are, respectively, e^{kx} for $x < 0$ and e^{-kx} for $x > 0$, with $k^2 = \frac{2m}{\hbar^2}|U_0 - E|$. Imposing the continuity of the wave function $\psi(x)$ and its derivative $\psi'(x)$ at $x = \pm\frac{a}{2}$, as well as the normalization condition $\int_{-\infty}^{\infty}|\psi(x)|^2 dx = 1$ one gets the transcendent equations for the eigenvalues of the even and odd solutions

$$\tan(ka) = \frac{\sqrt{C^2 - (ka)^2}}{ka},$$

respectively,

Fig. 2.6 Potential well

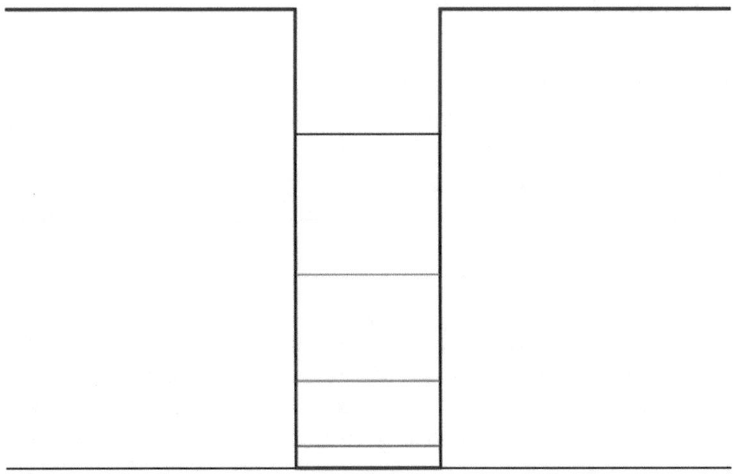

Fig. 2.7 Energy levels in the well for $C^2 = 38$

$$\tan(ka) = -\frac{ka}{\sqrt{C^2 - (ka)^2}},$$

where

$$C^2 = \frac{2ma^2}{\hbar^2} U_0.$$

The resulting energy spectrum consists of a finite number of discrete energies below U_0. For $C^2 = 38$ they are illustrated in Fig. 2.7. The shown levels correspond alternately to symmetric and anti-symmetric states. Of course, one has always a continuum above the barrier.

2.3 Electrons in a Periodical Potential

Periodical potentials in a crystal play a central role. The main properties of solids may be understood starting from the peculiarities of the quantum mechanical electron motion in a periodical potential.

2.3.1 Crystal Lattice

Most of the solid-state devices are made out of crystals and a theoretical treatment of periodical structures is much easier than that of disordered structures. In what follows we shall discuss only such solids with a periodical structure.

A crystal is characterized first of all by a translation symmetry of the (average) positions $\mathbf{R}$ of the atoms. Such a symmetry is defined by a 3D Bravais lattice

$$\mathbf{R} = n_1\mathbf{a}_1 + n_2\mathbf{a}_2 + n_3\mathbf{a}_3; \qquad (n_1, n_2, n_3 = 0, \pm 1, \pm 2, \dots).$$

The three vectors $\mathbf{a}_1, \mathbf{a}_2, \mathbf{a}_3$ define the elementary cell, whose repetition covers all the atoms. Its choice is not unique, but should contain the minimal possible number of atoms. An illustration of a Bravais lattice with a single atom in the elementary cell is shown in Fig. 2.8, for the sake of simplicity in 2D. An example of a more complicated elementary cell with two atoms is given in Fig. 2.9. One illustrates other possible choices of the elementary cell (see the arrows!). The volume of the elementary cell is given by

$$v = \mathbf{a}_1 \cdot (\mathbf{a}_2 \times \mathbf{a}_3)$$

and does not depend on the choice of the elementary cell. The elementary cell may contain different atoms and there are also crystals with elementary cells containing hundreds of atoms.

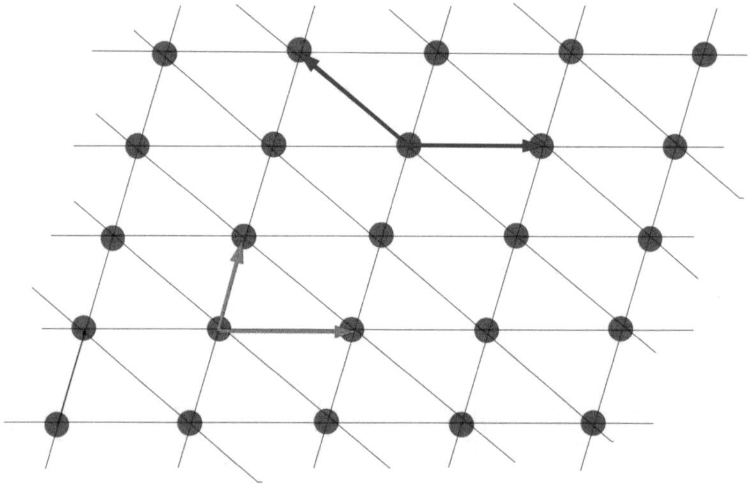

Fig. 2.8 Illustration of a crystal lattice in 2D, with two different choices of the elementary cell (indicated by the green, respectively, blue arrows)

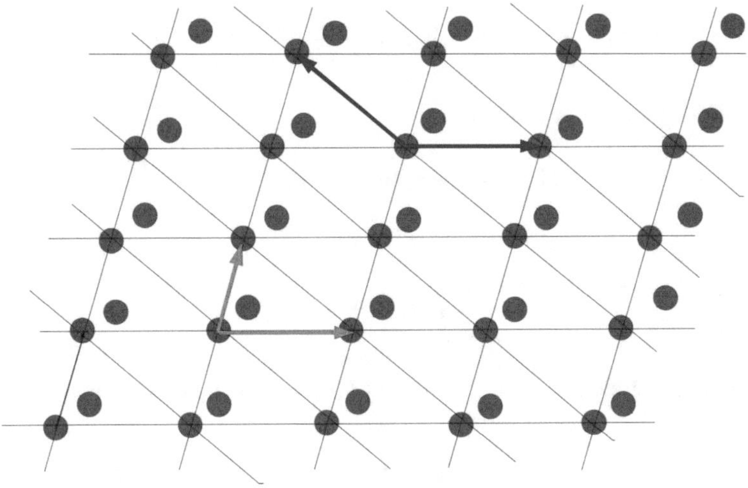

Fig. 2.9 Illustration of a Bravais lattice with an elementary cell of two atoms in 2D

The discrete translational symmetry is not the only possible symmetry of a crystal. It may be symmetric also with respect to certain rotations or mirrorings. In what follows we shall consider only the periodicity, which determines the basic properties of any crystal. The unit vectors $\mathbf{a}_1, \mathbf{a}_2, \mathbf{a}_3$ are not orthogonal to each other (except in a cubic crystal), therefore it is useful to introduce also a dual basis (called "reciprocal") by

$$\mathbf{b}_1 = \frac{2\pi}{v}\mathbf{a}_2 \times \mathbf{a}_3$$
$$\mathbf{b}_2 = \frac{2\pi}{v}\mathbf{a}_3 \times \mathbf{a}_1$$
$$\mathbf{b}_3 = \frac{2\pi}{v}\mathbf{a}_1 \times \mathbf{a}_2.$$

Then we have the orthogonality of the two dual unit vector sets:

$$\mathbf{b}_i \cdot \mathbf{a}_j = 2\pi\delta_{i,j}.$$

One may define also a reciprocal lattice created by the new basis vectors. A vector $\mathbf{K}$ of this reciprocal lattice is defined by

$$\mathbf{K} = m_1\mathbf{b}_1 + m_2\mathbf{b}_2 + m_3\mathbf{b}_3; \qquad (m_1, m_2, m_3 = 0, \pm 1, \pm 2, \dots).$$

The volume of the elementary cell of this lattice v_{BZ} called Brillouin zone (BZ) is related to that of the elementary cell of the original lattice v by

$$v_{BZ} = \mathbf{b}_1 \cdot (\mathbf{b}_2 \times \mathbf{b}_3) = \frac{(2\pi)^3}{\mathbf{a}_1 \cdot (\mathbf{a}_2 \times \mathbf{a}_3)} = \frac{(2\pi)^3}{v}.$$

The following relations also hold:

$$\mathbf{R} \cdot \mathbf{K} = 2\pi(n_1 m_1 + n_2 m_2 + n_3 m_3)$$

$$e^{\imath \mathbf{R} \cdot \mathbf{K}} = 1$$

$$\frac{1}{v} \int_v d\mathbf{r} e^{\imath \mathbf{r} \cdot \mathbf{K}} = \delta_{\mathbf{K},0}.$$

Any periodical function f on the lattice ($f(\mathbf{r}) = f(\mathbf{r} + \mathbf{R})$) may be expanded in Fourier series

$$f(\mathbf{r}) = \sum_{\mathbf{K}} \tilde{f}_{\mathbf{K}} e^{\imath \mathbf{r} \cdot \mathbf{K}}$$

with the Fourier coefficients

$$\tilde{f}_{\mathbf{K}} = \frac{1}{v} \int_v d\mathbf{r} f(\mathbf{r}) e^{-\imath \mathbf{r} \cdot \mathbf{K}}.$$

2.3.2 *Bloch Functions*

Let us now consider the Hamiltonian of an electron

$$H = -\frac{\hbar^2}{2m} \nabla^2 + U(\mathbf{r})$$

in a periodical potential

$$U(\mathbf{r}) = U(\mathbf{r} + \mathbf{R}).$$

Here we want to exploit just those general features that emerge from the periodicity. The stationary states are solutions of the equation

$$H\psi = E\psi.$$

Since a perfectly periodical crystal has an infinite extension, the stationary solutions (like the plane waves in free space) cannot be normalized, and in this sense, they are not true eigenstates. From the periodicity it follows that the shifted wave function $\psi(\mathbf{r} + \mathbf{R})$ must be a stationary state with the same energy E and (in the absence of degeneracy) it may differ from the unshifted one only by a phase factor

$$\psi_E(\mathbf{r}+\mathbf{R}) = e^{\imath\phi_\mathbf{R}}\psi_E(\mathbf{r}).$$

It follows immediately the additivity of the phase

$$\phi_{\mathbf{R}_1+\mathbf{R}_2} = \phi_{\mathbf{R}_1} + \phi_{\mathbf{R}_2}$$

and therefore one may write

$$\psi_E(\mathbf{r}+\mathbf{R}) = e^{\imath\mathbf{kR}}\psi_E(\mathbf{r})$$

with a real vector **k**. According to the previous definition of the reciprocal lattice, **k** and $\mathbf{k}+\mathbf{K}$ are equivalent, since the scalar product **RK** is a multiple of 2π. One may thus write the wave function in the form

$$\psi_\mathbf{k}(\mathbf{r}) = e^{\imath\mathbf{kr}}u_\mathbf{k}(\mathbf{r})$$

where $u_\mathbf{k}(\mathbf{r})$ is periodical

$$u_\mathbf{k}(\mathbf{r}+\mathbf{R}) = u_\mathbf{k}(\mathbf{r})$$

and the wave vector **k** belongs to the elementary cell of the reciprocal lattice. We choose it as the cell at the origin (BZ) defined as the domain

$$\mathbf{k}^2 \le (\mathbf{k}-\mathbf{K})^2$$

of **k**-vectors closer to the origin as any non-vanishing vector **K** of the reciprocal lattice. Even in the case of degeneracy one may choose the wave functions in this Bloch form. (This stems from the fact that the translations are an abelian group with one-dimensional irreducible representations.)

Thus, beside the energy we have the wave vector **k** for the characterization of the Bloch states. Of course, other quantum numbers must be considered, which we shall denote by n. These quantum numbers are discrete and are called—band indices.

2.3.3 Periodical Boundary Conditions

Just like in the case of the free electrons it is comfortable to impose some boundary conditions to get a discrete spectrum. In the case of a crystal lattice one chooses a periodical boundary condition compatible with the lattice periodicity

$$\psi(\mathbf{r}+\mathcal{N}_i\mathbf{a}_i) = \psi(\mathbf{r})$$

with some integer $\mathcal{N}_i \quad (i = 1, 2, 3)$. The parallelepiped with the edges $\mathcal{N}_i \mathbf{a}_i$ is called basic volume Ω. From the definition, it follows that the condition

$$\mathcal{N}_i \mathbf{a}_i \cdot \mathbf{k} = 2\pi m; \qquad (m = 0, \pm 1, \pm 2, \ldots)$$

has to be satisfied in order to recover the same phase factor of the Bloch function and therefore the wave vector **k** may take only discrete values

$$\mathbf{k} = \frac{n_1}{\mathcal{N}_1}\mathbf{b}_1 + \frac{n_2}{\mathcal{N}_2}\mathbf{b}_2 + \frac{n_3}{\mathcal{N}_3}\mathbf{b}_3; \qquad (n_1, n_2, n_3 = 0, \pm 1, \pm 2, \ldots).$$

Of course, to obtain physically relevant results for a bulk crystal one has to remove the dependence on these artificial boundary conditions by taking the infinite volume limit. One sees that as $\mathcal{N}_i (i = 1, 2, 3)$ increases, the elementary step of the wave vectors **k** goes to zero, therefore in the limit of infinite basic volume in all directions one has

$$\frac{1}{\mathcal{N}_1 \mathcal{N}_2 \mathcal{N}_3 v} \sum_{\mathbf{k} \in BZ} \longrightarrow \frac{1}{(2\pi)^3} \int_{BZ} d^3\mathbf{k}.$$

The introduction of the periodical boundary conditions allows the orthonormalization of the Bloch functions

$$\int_{\Omega} d\mathbf{r} \psi_{\mathbf{k},n}(\mathbf{r})^* \psi_{\mathbf{k}',n'}(\mathbf{r}) = \delta_{\mathbf{k},\mathbf{k}'} \delta_{n,n'}.$$

Now, the integration over the basic volume Ω is equivalent to a summation over all the elementary cells and the integration over each cell. With

$$\psi_{\mathbf{k},n}(\mathbf{r}) = \frac{1}{\sqrt{\mathcal{N}}} e^{\imath \mathbf{k}\mathbf{r}} u_{\mathbf{k},n}; \qquad \mathcal{N} \equiv \mathcal{N}_1 \mathcal{N}_2 \mathcal{N}_3$$

the orthonormality may be written also as

$$\frac{1}{\mathcal{N}} \sum_{\mathbf{R} \in \Omega} e^{\imath(\mathbf{k}-\mathbf{k}')\mathbf{R}} \int_v d\mathbf{r} e^{\imath(\mathbf{k}-\mathbf{k}')\mathbf{r}} u_{\mathbf{k},n}(\mathbf{r})^* u_{\mathbf{k}',n'}(\mathbf{r}) = \delta_{\mathbf{k},\mathbf{k}'} \int_v d\mathbf{r} u_{\mathbf{k},n}(\mathbf{r})^* u_{\mathbf{k}',n'}(\mathbf{r}),$$

where we used the periodicity of the Bloch part and the relation

$$\frac{1}{\mathcal{N}} \sum_{\mathbf{R} \in \Omega} e^{\imath \mathbf{k}\mathbf{R}} = \delta_{\mathbf{k},0}$$

stemming from the identity

$$\frac{1}{\mathscr{M}}\sum_{n=0}^{\mathscr{M}-1} x^n = \frac{1}{\mathscr{M}}\frac{1-x^{\mathscr{M}}}{1-x} = \begin{cases} 0\ if\ m \neq 0 \\ 1\ if\ m = 0 \end{cases},$$

with $x \equiv e^{\imath\frac{2\pi m}{\mathscr{M}}}$; $\quad (m = 0, \ldots, \mathscr{M}-1)$. Thus,

$$\int_v d\mathbf{r} u_{\mathbf{k},n}(\mathbf{r})^* u_{\mathbf{k},n'}(\mathbf{r}) = \delta_{n,n'}$$

and therefore the Bloch parts $u_{\mathbf{k},n}(\mathbf{r})$ form for each $\mathbf{k}$ an orthonormalized set and at any fixed $\mathbf{k}$ one has a discrete spectrum.

2.3.4 The Approximation of Quasi-Free Electrons

Let us write the Schrödinger eigenvalue equation in the periodical potential $U(\mathbf{r})$ in Fourier components

$$u_{\mathbf{k}}(\mathbf{r}) = \sum_{\mathbf{K}} \tilde{u}_{\mathbf{k},\mathbf{K}} e^{\imath\mathbf{K}\cdot\mathbf{r}}$$

$$U(\mathbf{r}) = \sum_{\mathbf{K}} \tilde{U}_{\mathbf{K}} e^{\imath\mathbf{K}\cdot\mathbf{r}},$$

where $\tilde{U}_{\mathbf{K}} = \tilde{U}^*_{-\mathbf{K}}$ follows from reality of the potential and the wave vector $\mathbf{k}$ here is restricted to the Brillouin zone. After multiplication with $e^{-\imath\mathbf{K}\cdot\mathbf{r}}$ and integration of $\mathbf{r}$ over the elementary cell the equation looks as

$$\left[\frac{\hbar^2}{2m}(\mathbf{k}+\mathbf{K})^2 - E_{\mathbf{K}}(\mathbf{k})\right]\tilde{u}_{\mathbf{k},\mathbf{K}} + \sum_{\mathbf{K}'} \tilde{U}_{\mathbf{K}-\mathbf{K}'}\tilde{u}_{\mathbf{k},\mathbf{K}'} = 0.$$

Let us consider first an empty lattice $U(\mathbf{r}) = 0$. The corresponding states and eigenenergies are

$$\psi^{(0)}_{\mathbf{k},\mathbf{K}}(\mathbf{r}) = \frac{1}{\sqrt{\mathscr{N}}} e^{\imath\mathbf{k}\mathbf{r}} u^{(0)}_{\mathbf{k},\mathbf{K}}(\mathbf{r}), \qquad u^{(0)}_{\mathbf{k},\mathbf{K}}(\mathbf{r}) = \frac{1}{\sqrt{v}} e^{\imath\mathbf{K}\mathbf{r}},$$

respectively,

$$E^{(0)}_{\mathbf{K}}(\mathbf{k}) = \frac{\hbar^2}{2m}(\mathbf{k}+\mathbf{K})^2; \qquad \mathbf{k} \in BZ.$$

The "bands" indicated by the vectors $\mathbf{K}$ of the reciprocal lattice are shown in Fig. 2.10 for $K = -2\pi, 0, 2\pi$, (with lattice constant $a = 1$) in the case of an one-dimensional lattice.

It is important to remark again that at fixed k the spectrum is discrete. One sees the degeneracy of the "bands" at $k = \pm\pi$ and at $k = 0$. The main role of the perturbation by a small periodical potential $U(\mathbf{r})$ will be in lifting these degeneracies. As we shall see on an exactly soluble one-dimensional example, after the lifting of the degeneracy the top of the lowest band and the bottom of the upper band get extrema with vanishing derivative.

2.3.5 The Kronig–Penney Model

We consider here explicitly the solvable one-dimensional periodical potential model (see Fig. 2.11)

$$V(x) = \begin{cases} 0 \quad n(a+b) < x < n(a+b)+a; & n = 0, \pm 1, \pm 2, \ldots \\ U_0 \quad n(a+b) - b \leq x \leq n(a+b). \end{cases}$$

Actually one has to solve the problem in the elementary cell ($0 < x < l, \equiv a+b$) and thereafter extend it overall by the Bloch condition. For $E < U_0$ the solution is

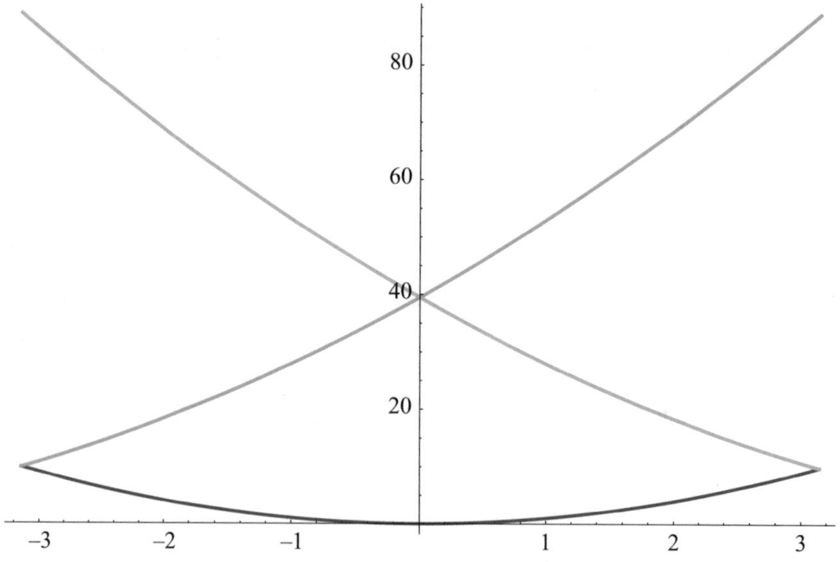

Fig. 2.10 The first folded free electron bands in 1D

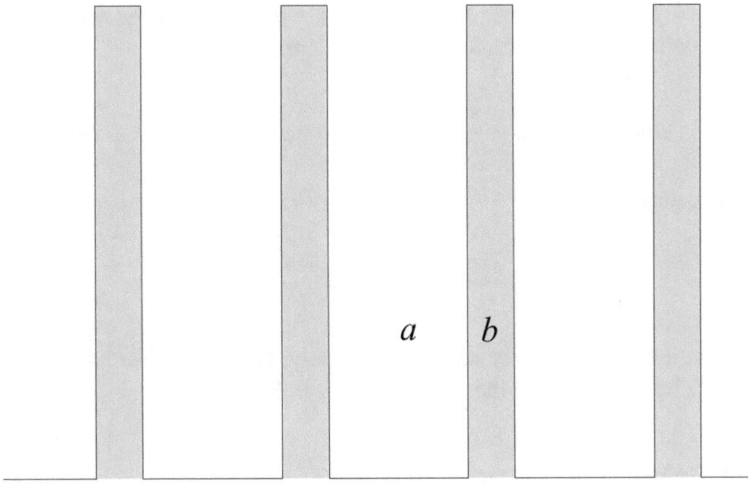

Fig. 2.11 The Kronig–Penney periodic potential

$$\Psi(x) = c_1^{II} e^{i\kappa_2 x} + c_2^{II} e^{-i\kappa_2 x}; \qquad 0 < x < a$$
$$\Psi(x) = c_1^{I} e^{\kappa_1 x} + c_2^{I} e^{-\kappa_1 x}; \qquad a < x < l$$

$$\kappa_1^2 = \frac{2m}{\hbar^2} E, \qquad \kappa_2^2 = \frac{2m}{\hbar^2} |U_0 - E|.$$

At first one must ensure the continuity of the wave function and its derivative at $x = a$

$$\Psi(a-0) = \Psi(a+0)$$
$$\Psi'(a-0) = \Psi'(a+0).$$

Further one has still to fulfill the Bloch condition

$$\Psi(0)e^{ikl} = \Psi(l)$$
$$\Psi'(0)e^{ikl} = \Psi'(l)\,.$$

These conditions lead to a system of four linear homogeneous equations for the four coefficients c_1^I, c_1^{II}, c_2^I, c_2^{II}. The existence of a solution implies a transcendent equation for the energy as function of the wave vector k (energy bands). For $E < U_0$ it looks as

$$\cosh\kappa_2 b \cos\kappa_1 a + \frac{\kappa_2^2 - \kappa_1^2}{2\kappa_1\kappa_2} \sinh\kappa_2 b \sin\kappa_1 a = \cos k(a+b).$$

For $E > U_0$ trigonometric functions are the solution and the above discussed boundary conditions lead then to the transcendent equation

$$\cos \kappa_1 b \cos \kappa_2 a - \frac{\kappa_1^2 + \kappa_2^2}{2\kappa_1 \kappa_2} \sin \kappa_1 b \sin \kappa_2 a = \cos k(a+b)$$

where, however,

$$\kappa_1^2 = \frac{2m}{\hbar^2} E, \qquad \kappa_2^2 = \frac{2m}{\hbar^2} |E - U_0|.$$

These equations may be easily solved numerically. The lowest four bands are shown in Fig. 2.12 by the choice of the parameters $U_0 = 38, b = 0.25a$.

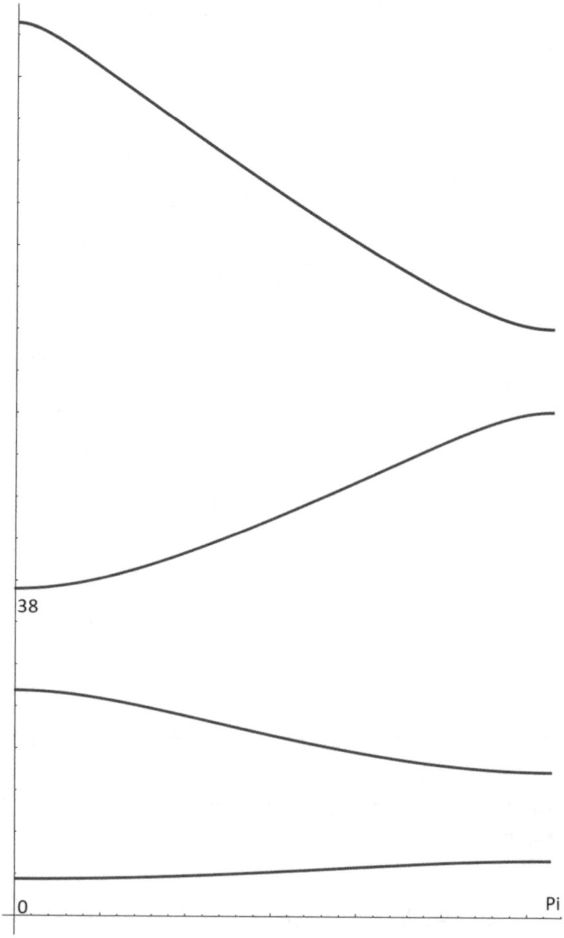

Fig. 2.12 The lowest four energy bands (here $\frac{2ma^2}{\hbar^2} E$) of the Kronig–Penney model with $b = 0.25a$, $\frac{2ma^2}{\hbar^2} U_0 = 38$ for $0 < k(a+b) < \pi$

Fig. 2.13 The lowest four energy bands (here $\frac{2ma^2}{\hbar^2}E$) of the Kronig–Penney model with $b = a$, $\frac{2ma^2}{\hbar^2}U_0 = 38$ for $0 < k(a+b) < \pi$. (Bands below U_0 here in red)

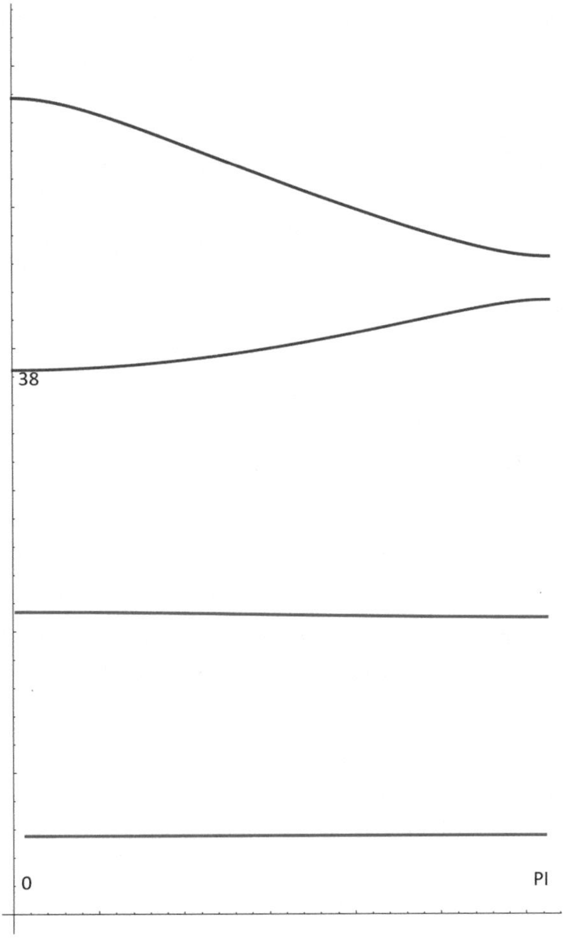

One sees here the scenario described in the previous section. The bands of the empty lattice are split, and one gets extrema with vanishing derivatives.

It is instructive to remark here that as the distance between the wells b increases the bands under U_0 flatten (in our case the lowest two) and get close to the energy levels of the isolated well, while the upper bands get closer to those of the empty lattice. This is illustrated in Fig. 2.13 with the increased distance $b = a$ between the wells. (The range of the wave vector is limited here for the sake of convenience to $0 < k(a+b) < \pi$.) This result corresponds to the known fact that far away atoms have the spectrum of the isolated atoms, however, with a degeneracy equaling the total number of atoms.

2.3.6 *Band Extrema, kp: Perturbation Theory and Effective Mass*

The stationary Schrödinger equation for the Bloch functions describing the energy bands (indexed by some quantum numbers n)

$$\left[-\frac{\hbar^2}{2m}\nabla^2 + U(\mathbf{r}) - E_n(\mathbf{k})\right]\psi_{n,\mathbf{k}}(\mathbf{r}) = 0$$

after inserting

$$\psi_{n,\mathbf{k}}(\mathbf{r}) = e^{\imath\mathbf{kr}}u_{n,\mathbf{k}}(\mathbf{r});$$

gives rise to the Bloch equation

$$\left[-\frac{\hbar^2}{2m}\nabla^2 + U(\mathbf{r}) - \frac{\imath\hbar^2}{m}\mathbf{k}\nabla + \frac{\hbar^2}{2m}k^2 - E_n(\mathbf{k})\right]u_{\mathbf{k},n}(\mathbf{r}) = 0\,.$$

For any fixed $\mathbf{k}$ this is a true eigenvalue equation with discrete eigenvalues. The eigenfunctions $u_{\mathbf{k},n}(\mathbf{r})$ due to the Bloch property

$$u_{n,\mathbf{k}}(\mathbf{r}+\mathbf{R}) = u_{n,\mathbf{k}}(\mathbf{r})$$

may be considered just in the elementary cell with periodical boundary conditions. Therefore, they may form a complete orthonormalized set in the cell

$$\int_v d\mathbf{r}u_{n,\mathbf{k}}(\mathbf{r})^* u_{n',\mathbf{k}}(\mathbf{r}) = \delta_{n,n'}$$

$$\sum_n u_{n,\mathbf{k}}(\mathbf{r})u_{n,\mathbf{k}}(\mathbf{r}')^* = \delta(\mathbf{r},\mathbf{r}')\,.$$

Let us assume that for a given $\mathbf{k}_0$ the eigenfunctions $u_{n,\mathbf{k}_0}(\mathbf{r})$ and eigenenergies $E_n(\mathbf{k}_0)$ are known.

For a small deviation $|\mathbf{k}-\mathbf{k}_0| \to 0$, one may consider $-\frac{\imath\hbar^2}{m}(\mathbf{k}-\mathbf{k}_0)\nabla$ as a small perturbation. In the absence of degeneracy, standard perturbation theory gives

$$u_{n,\mathbf{k}}(\mathbf{r}) = u_{n,\mathbf{k}_0}(\mathbf{r}) + \sum_{n'\neq n}\frac{< n',\mathbf{k}_0|\frac{\hbar^2(\mathbf{k}-\mathbf{k}_0)\nabla}{\imath m}|n,\mathbf{k}_0 >}{E_n(\mathbf{k}_0) - E_{n'}(\mathbf{k}_0)}u_{n',\mathbf{k}_0}(\mathbf{r}) + \cdots$$

and

$$E_n(\mathbf{k}) - \frac{\hbar^2}{2m}k^2 = E_n(\mathbf{k}_0) - \frac{\hbar^2}{2m}k_0^2 + < n, \mathbf{k}_0|\frac{\hbar(\mathbf{k}-\mathbf{k}_0)}{\imath m}\nabla|n, \mathbf{k}_0 >$$

$$+ \sum_{n' \neq n} \frac{\left| < n, \mathbf{k}_0|\frac{\hbar^2(\mathbf{k}-\mathbf{k}_0)\nabla}{\imath m}|n', \mathbf{k}_0 > \right|^2}{E_n(\mathbf{k}_0) - E_{n'}(\mathbf{k}_0)} + \cdots$$

with the "bra-ket" notation

$$< n, \mathbf{k}_0|\nabla|n', \mathbf{k}_0 > \equiv \int_v d\mathbf{r} u_{n,\mathbf{k}_0}(\mathbf{r})^* \nabla u_{n',\mathbf{k}_0}(\mathbf{r}) .$$

It follows then for the derivative of the energy at $\mathbf{k} = \mathbf{k}_0$

$$\frac{\partial E_n(\mathbf{k})}{\partial \mathbf{k}}|_{\mathbf{k}=\mathbf{k}_0} = < n, \mathbf{k}_0|\frac{\hbar^2}{\imath m}\nabla|n, \mathbf{k}_0 > .$$

Since $\mathbf{k}_0$ for the time being is arbitrary, the relation holds for any $\mathbf{k}$ and we get for the average velocity

$$\langle \mathbf{v} \rangle \equiv \int_v d\mathbf{r} \psi_{n,\mathbf{k}}(\mathbf{r})^* \frac{\hbar}{\imath m} \nabla \psi_{n,\mathbf{k}}(\mathbf{r}) = \frac{1}{\hbar}\frac{\partial E_n(\mathbf{k})}{\partial \mathbf{k}} .$$

Now let us assume that the point $\mathbf{k}_0$ is a band extremum. Then the linear terms in the energy vanish and

$$E_n(\mathbf{k}) = E_n(\mathbf{k}_0) + \frac{\hbar^2}{2}\left(\frac{1}{\mathcal{M}}\right)_{\mu\nu} (k-k_0)_\mu (k-k_0)_\nu + \cdots ,$$

where

$$\left(\frac{1}{\mathcal{M}}\right)_{\mu\nu} = \frac{1}{m}\delta_{\mu\nu} + \frac{2}{m^2}\sum_{n' \neq n} \frac{< n, 0|\hbar\nabla_\mu|n', 0 >< n', 0|\hbar\nabla_\nu|n, 0 >}{E_n(0) - E_{n'}(0)} + \cdots$$

is the *inverse effective mass tensor.* In the case of the cubic symmetry with the extremum at $\mathbf{k}_0 = 0$ this looks as

$$E_n(\mathbf{k}) = E_n(0) + \frac{\hbar^2}{2m^*}k^2 + \cdots$$

and one has an isotropic *effective mass* m^*. This effective mass may be positive or negative, depending on the nature of the extremum (maximum or minimum). Therefore, depending on the sign of the effective mass, the average velocity

$$\langle \mathbf{v} \rangle = \frac{\hbar}{m^*}\mathbf{k}$$

may be either in the direction of the wave vector **k** or opposite to it.

We have seen on the example of the one-dimensional Kronig–Penney model that band extrema occur at $k = 0, \pm\frac{\pi}{a}$. Most of the important properties of crystals are determined by the extrema of some of the energy bands we shall discuss later.

2.3.7 *Wannier Functions and Tight-Binding Approximation*

A useful concept in the treatment of motion in a periodical potential are the Wannier functions defined through the Bloch functions of a given band n by

$$w_{n\mathbf{R}}(\mathbf{r}) \equiv \frac{1}{\sqrt{\mathcal{N}}} \sum_{\mathbf{k}\in BZ} \psi_{n,\mathbf{k}}(\mathbf{r}) e^{-\imath\mathbf{k}\mathbf{R}} = \frac{1}{\sqrt{\mathcal{N}}} \sum_{\mathbf{k}\in BZ} u_{n,\mathbf{k}}(\mathbf{r}) e^{\imath\mathbf{k}(\mathbf{r}-\mathbf{R})} .$$

They constitute a complete orthonormalized system of functions

$$\int_{\Omega} d\mathbf{r} w_{n\mathbf{R}}(\mathbf{r})^* w_{n'\mathbf{R}'}(\mathbf{r}) = \frac{1}{\mathcal{N}} \sum_{\mathbf{k},\mathbf{k}'\in BZ} e^{\imath(\mathbf{k}\mathbf{R}-\mathbf{k}'\mathbf{R}')} \int_{v} d\mathbf{r} \psi_{n,\mathbf{k}}(\mathbf{r})^* \psi_{n',\mathbf{k}'}(\mathbf{r})$$

$$= \frac{1}{\mathcal{N}} \sum_{\mathbf{k}\in BZ} e^{\imath\mathbf{k}(\mathbf{R}-\mathbf{R}')} \delta_{n,n'} = \delta_{n,n'} \delta_{\mathbf{R},\mathbf{R}'}$$

and

$$\sum_{n,\mathbf{R}} w_{n\mathbf{R}}(\mathbf{r})^* w_{n\mathbf{R}}(\mathbf{r}') = \delta(\mathbf{r}, \mathbf{r}').$$

The inverse transformation is

$$u_{n,\mathbf{k}}(\mathbf{r}) = \frac{1}{\sqrt{\mathcal{N}}} \sum_{\mathbf{R}} e^{\imath\mathbf{k}(\mathbf{r}-\mathbf{R})} w_{n,\mathbf{R}}(\mathbf{r}).$$

The Wannier functions, however, are not eigenfunctions of the Hamiltonian. If $u_{n,\mathbf{k}}(\mathbf{r})$ is a smooth function of **k**, then $w_{n\mathbf{R}}(\mathbf{r})$ vanishes rapidly for $|\mathbf{r} - \mathbf{R}| \to \infty$. Therefore, while the Bloch eigenfunctions are delocalized, the Wannier functions are localized on the lattice nodes. Since the Wannier functions are constructed only from the Bloch functions of a single band and considering also the discrete translational invariance on the lattice, the matrix elements of the Hamiltonian between the Wannier functions depend only on the vector $\mathbf{R} - \mathbf{R}'$

$$< n, \mathbf{R}|H|n', \mathbf{R}' >= \delta_{n,n'} t_n(\mathbf{R} - \mathbf{R}') .$$

Therefore,

$$\begin{aligned}
< n, \mathbf{R}|H|n, \mathbf{k} > &= \frac{1}{\sqrt{\mathcal{N}}} \sum_{\mathbf{R}'} < n, \mathbf{R}|H|n, \mathbf{R}' > e^{-\imath \mathbf{k}\mathbf{R}'} \\
&= \frac{1}{\sqrt{\mathcal{N}}} \sum_{\mathbf{R}'} t_n(\mathbf{R} - \mathbf{R}') e^{\imath \mathbf{k}\mathbf{R}'} \\
&= e^{\imath \mathbf{k}\mathbf{R}} \frac{1}{\sqrt{\mathcal{N}}} \sum_{\mathbf{R}'} t_n(\mathbf{R}') e^{\imath \mathbf{k}\mathbf{R}'} .
\end{aligned}$$

On the other hand

$$\begin{aligned}
< n, \mathbf{R}|H|n, \mathbf{k} > &= E_n(\mathbf{k}) < n, \mathbf{R}|n, \mathbf{k} > \\
&= E_n(\mathbf{k}) \frac{1}{\sqrt{\mathcal{N}}} \sum_{\mathbf{R}'} < n, \mathbf{R}|n, \mathbf{R}' > e^{\imath \mathbf{k}\mathbf{R}'} . \\
&= e^{\imath \mathbf{k}\mathbf{R}} E_n(\mathbf{k}) \frac{1}{\sqrt{\mathcal{N}}}
\end{aligned}$$

and therefore, inverting the relation we get

$$E_n(\mathbf{k}) = \sum_{\mathbf{R}} e^{\imath \mathbf{k}\mathbf{R}} t_n(\mathbf{R}) .$$

As an illustration, let us assume that in a cubic lattice of lattice constant a the only non-vanishing $t_n(\mathbf{R})$ are those to the nearest neighbors

$$t_n(0) = c_n , \quad t_n(\pm \mathbf{a}_x) = t_n(\pm \mathbf{a}_y) = t_n(\pm \mathbf{a}_z) = -d_n .$$

Then it follows:

$$E_n(\mathbf{k}) = c_n - 2d_n(\cos(k_x a) + \cos(k_y a) + \cos(k_z a))$$

and in the vicinity of the origin

$$E_n(\mathbf{k}) = c_n - 6d_n + d_n a^2 k^2 + \dots$$

This defines the effective mass by $\frac{\hbar^2}{2m^*} = d_n a^2$.

One approximates often the Wannier functions by the eigenfunctions of the isolated atom $\phi_n(\mathbf{r})$ (“tight-binding approximation”)

$$w_{n\mathbf{R}}(\mathbf{r}) \approx \phi_n(\mathbf{r} - \mathbf{R}) .$$

Within this approximation

$$\psi_{n,\mathbf{k}}(\mathbf{r}) = \frac{1}{\sqrt{\mathcal{N}}} \sum_{\mathbf{R}} e^{-\imath \mathbf{k}\mathbf{R}} \phi_n(\mathbf{r}-\mathbf{R})$$

or

$$u_{n,\mathbf{k}}(\mathbf{r}) = \frac{1}{\sqrt{\mathcal{N}}} \sum_{\mathbf{R}} e^{\imath \mathbf{k}(\mathbf{r}-\mathbf{R})} \phi_n(\mathbf{r}-\mathbf{R})\,.$$

It is, however, important to remark that the orthonormalization of these functions is not fulfilled since $\int_{\Omega} d\mathbf{r}\phi_n(\mathbf{r})^* \phi_{n'}(\mathbf{r}-\mathbf{R})$, however, small, it does not vanish for $\mathbf{R} \neq 0$. By this connection to the atomic wave functions one may speak about a given band as characterized by a given symmetry of the underlying atomic states. For example, s or p-like states, with their respective degeneracies.

2.3.8 *Bloch Electron in a Homogeneous Electric Field*

Let us consider a homogeneous electric field superimposed on the periodical potential. Thus, we have a supplementary term $H' = -e\mathbf{E}\mathbf{r}$ in the Hamiltonian. The resulting potential looks like in Fig. 2.14. Let us choose the normalization of the Bloch functions as

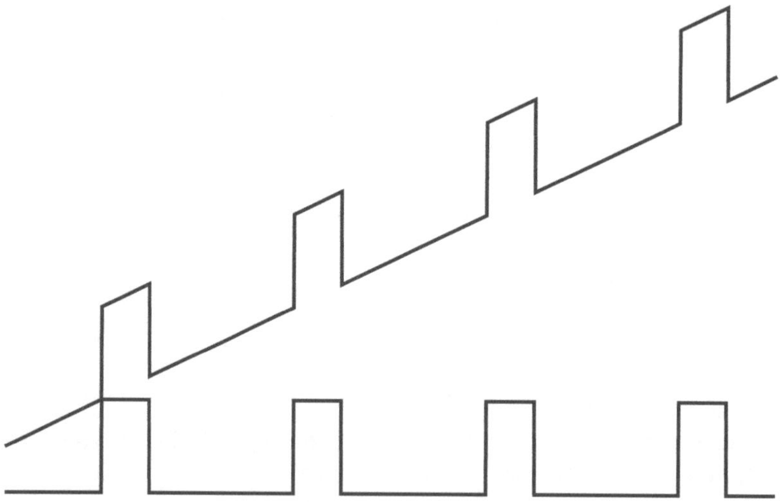

Fig. 2.14 A periodic potential and its modification due to a homogeneous electric field

$$< \mathbf{k}', n'|\mathbf{k}, n >= v_{BZ}^{-1}\delta(\mathbf{k}-\mathbf{k}')\delta_{n,n'}.$$

Since in the infinite volume limit $\mathcal{N}\delta_{\mathbf{k},\mathbf{k}'} \to v_{BZ}\delta(\mathbf{k}-\mathbf{k}')$, this normalization corresponds in the discretized version to $< \mathbf{k}', n'|\mathbf{k}, n >= N\delta_{\mathbf{k},\mathbf{k}'}\delta_{n,n'}$. Then

$$\begin{aligned}
&\int d\mathbf{r}\psi_{\mathbf{k}',n'}(\mathbf{r})^*\mathbf{r}\psi_{\mathbf{k},n}(\mathbf{r}) = \int d\mathbf{r}e^{\imath(\mathbf{k}-\mathbf{k}')\mathbf{r}}\mathbf{r}u_{\mathbf{k}',n'}(\mathbf{r})^*u_{\mathbf{k}n}(\mathbf{r}) \\
&= -\imath\tfrac{\partial}{\partial\mathbf{k}}\int d\mathbf{r}\psi_{\mathbf{k}',n'}(\mathbf{r})^*\psi_{\mathbf{k},n}(\mathbf{r}) + \imath\int d\mathbf{r}e^{\imath(\mathbf{k}-\mathbf{k}')\mathbf{r}}u_{\mathbf{k}',n'}(\mathbf{r})^*\tfrac{\partial}{\partial\mathbf{k}}u_{\mathbf{k},n}(\mathbf{r}) \\
&= -\imath v_{BZ}^{-1}\tfrac{\partial}{\partial\mathbf{k}}\delta(\mathbf{k}-\mathbf{k}')\delta_{n,n'} + \imath\sum_{\mathbf{R}} e^{\imath(\mathbf{k}-\mathbf{k}')\mathbf{R}}\int_v d\mathbf{r}e^{\imath(\mathbf{k}-\mathbf{k}')\mathbf{r}}u_{\mathbf{k}',n'}(\mathbf{r})^*\tfrac{\partial}{\partial\mathbf{k}}u_{\mathbf{k},n}(\mathbf{r}) \\
&= -\imath v_{BZ}^{-1}\tfrac{\partial}{\partial\mathbf{k}}\delta(\mathbf{k}-\mathbf{k}')\delta_{n,n'} + \imath v_{BZ}\delta(\mathbf{k}-\mathbf{k}')\int_v d\mathbf{r}u_{\mathbf{k},n'}(\mathbf{r})^*\tfrac{\partial}{\partial\mathbf{k}}u_{\mathbf{k},n}(\mathbf{r}).
\end{aligned}$$

If one ignores inter-band matrix elements as well as the intra-band energy correction, we get the approximate expression for the matrix elements of H':

$$< \mathbf{k}', n'|H'|\mathbf{k}n >\approx -\imath v_{BZ}^{-1}e\mathbf{E}\frac{\partial}{\partial\mathbf{k}}\delta(\mathbf{k}-\mathbf{k}')\delta_{n.n'}.$$

Consequently, we obtain in the $\mathbf{k}$ space an effective one-band Hamiltonian (in each band)

$$H'_{eff} \equiv E(\mathbf{k}) - \imath e\mathbf{E}\frac{\partial}{\partial\mathbf{k}}$$

and in the same representation the following Heisenberg equation of motion for the $\mathbf{k}$-vector

$$\hbar\dot{\mathbf{k}} = e\mathbf{E}$$

results.

Due to the periodicity of the band energy $E(\mathbf{k})$, after a time $\mathscr{T} = \frac{2\pi\hbar}{ea\mathscr{E}}$ the electron crosses the whole Brillouin zone and will be reflected at its boundary. Therefore, unlike the accelerated motion of a free electron, the electron in a crystal performs Bloch oscillations with the frequency $\omega = \frac{2\pi}{\mathscr{T}}$. However, these oscillations in a real crystal are not observable since due to the smallness of the lattice constant the Bloch period $\mathscr{T}$ is much bigger than the relaxation time due to the interaction with other perturbations of the lattice (phonons, impurities). The Bloch period itself cannot be shortened by the increase of the field strength due to the here neglected inter-band transitions, which may become important in strong electric fields. Nevertheless, Bloch oscillation may be observed in artificial periodical semiconductor structures with large lattice constants.

As in the case of the free electrons in a homogeneous field we may reinterpret these results in terms of stationary states (Wannier–Ladders). The stationary Schrödinger equation within our approximations looks as

$$\left[E(\mathbf{k}) - \imath e\mathbf{E}\frac{\partial}{\partial \mathbf{k}} - \epsilon\right]\tilde{\psi}(\mathbf{k}) = 0$$

or

$$\frac{\mathbf{E}}{\mathscr{E}}\frac{\partial \ln \tilde{\psi}(\mathbf{k})}{\partial \mathbf{k}} = \frac{\imath}{e\mathscr{E}}(\epsilon - E(\mathbf{k})).$$

Choosing the field along the x-axis we get

$$\tilde{\psi}(\mathbf{k}) = \tilde{\psi}(0, \mathbf{k}_\perp)\exp\left\{\frac{\imath}{e\mathscr{E}}\int_0^{k_x} dq_x\,(\epsilon - E(q_x, \mathbf{k}_\perp))\right\}.$$

We must still implement the periodic boundary condition for the wave function in the **k** space. Taking into account the periodicity of the band energy $E(\mathbf{k})$ (implying the vanishing of its integral over the Brillouin zone). This is achieved by fixing the eigenenergies

$$\epsilon_m(\mathbf{k}_\perp) = ea\mathscr{E}m + const. \qquad (m = 0, \pm 1, \pm 2, \ldots).$$

Then the difference between two successive energy levels is

$$\epsilon_{m+1}(\mathbf{k}_\perp) - \epsilon_m(\mathbf{k}_\perp) = ea\mathscr{E}.$$

It corresponds to the energy associated to the Bloch frequency $\omega = \frac{2\pi}{\mathscr{T}}$.

As an example, let us consider a simple one-dimensional Bloch spectrum $E(k) = \Delta(1 - \cos(ka))$. Then $\epsilon_m = ea\mathscr{E}m + \Delta$ and the wave function in real space (illustrated in Fig. 2.15 for $m = 0$) is

$$\psi_m(x) = const. \int_{-\frac{\pi}{a}}^{\frac{\pi}{a}} dk \exp\left\{-\imath k(x - ma) - \frac{\imath\Delta}{e\mathscr{E}}\sin(ka)\right\}.$$

2.4 Electronic Occupation of States in a Crystal

The occupation of the one-electron states in the periodical potential by as many electrons as ions at low temperatures determines the nature of the solid.

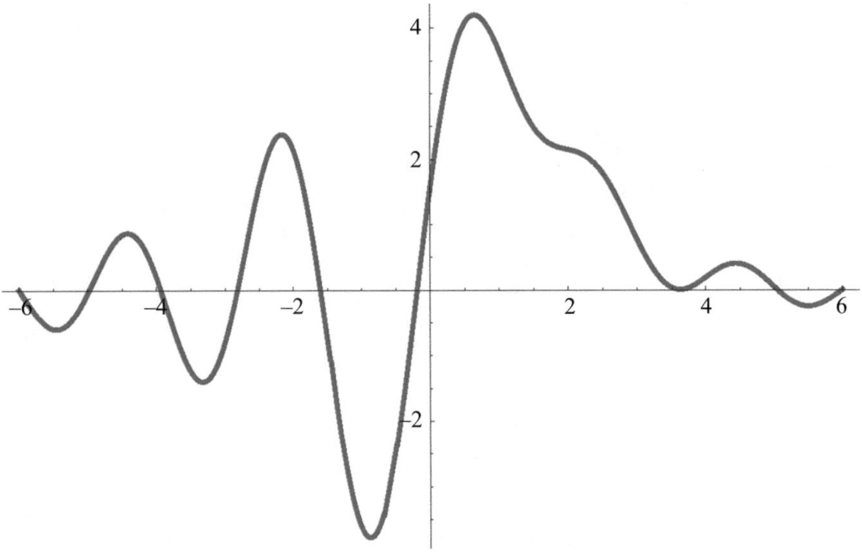

Fig. 2.15 Stationary state of the Wannier ladder

2.4.1 *Ground State Occupation of Bands: Conductors and Insulators*

We have discussed until now the nature of states of an electron in a crystal. However, the properties of the material are essentially determined by the occupation of these states. According to their electric conductance one classifies the crystals as metals or insulators. On the basis of the discussed one-electron spectra of crystals one may already understand the basic difference. The ground state of the electrons ($T = 0$) in a crystal results from the successive occupation of the lowest one-electron states in the bands. In this ground state corresponding to zero temperatūre, the highest band may be either partially or completely filled. In the first case, a small external perturbation can excite the electrons, while in the latter case, due to the forbidden states above them (band gap), only a strong perturbation may excite them. The partially filled band is called conduction band, while the last completely filled band is called valence band. The first case defines a conductor (metal), while the second defines an insulator (dielectric). The two situations are illustrated in the simple case of a direct band gap, when both the maximum of the valence band and the minimum of the conduction band lie at $\mathbf{k} = 0$ on Fig. 2.16. The red color indicates the occupied states in the bands.

However, there is a very important sort of dielectrics, the semiconductors. They behave as insulators at very low temperatures, while at higher temperatures, they behave as conductors. In this case the band gap is so small that at room temperatures already a lot of electrons are thermally excited into the conduction

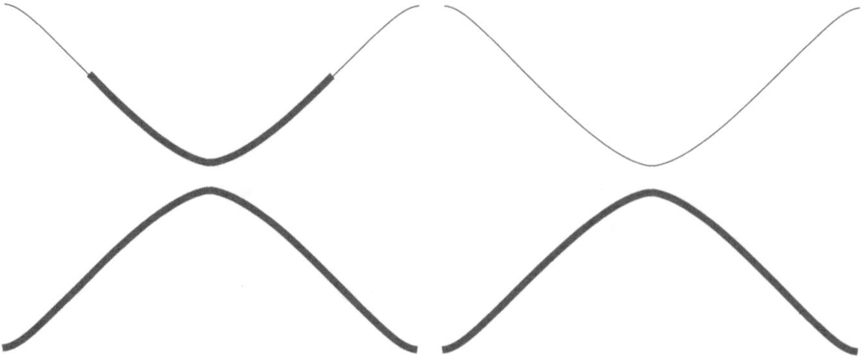

Fig. 2.16 Band filling in metals and insulators

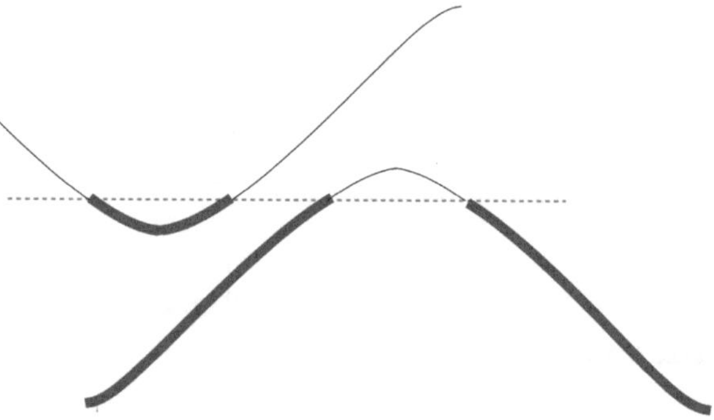

Fig. 2.17 Band filling in a semi-metal

band. In the presence of impurity states in the band gap, we shall discuss later, the number of excited electrons in the conduction band is even more important, and the conductivity may be relevant even at low temperatures.

Beside the above discussed cases, by crystals with indirect gaps, where the maximum of the valence band and the minimum of the conduction band occur at different values of **k**, it may happen that the later lies higher than the former. This band occupation at $T = 0$ of a semi-metal is schematically shown in Fig. 2.17.

2.4.2 Spin–Orbit Coupling and Valence Band Splitting

The discussion of the band structure of crystals is treated usually within the frame of the non-relativistic Schrödinger equation. However, there are often important corrections to be made due to relativistic effects. Most of the so-called direct gap semiconductors, having the maximum of the valence band and the minimum of the conduction band at $\mathbf{k} = 0$, within this frame get a non-degenerate conduction band and a threefold degenerate valence band. In a tight-binding approximation these may be considered as made up of s, respectively, p atomic states. Of course, one has to take into account also the spin degeneracy. However, actually the valence band states are split due to relativistic effect of the spin–orbit interaction. This interaction is invariant against simultaneous rotation of the orbital angular momentum $\mathbf{l}$ and of the spin $\boldsymbol{\sigma}$. Therefore, one has to classify the valence band states according to the states of the total angular momentum $\mathbf{J} \equiv \mathbf{l}+\boldsymbol{\sigma}$. From the $m = 0, \pm 1$ orbital angular momentum states and the $\sigma_z = \pm\frac{1}{2}$ spin states one has to construct the eigenstates of the total angular momentum $J = \frac{3}{2},\ J_z = \pm\frac{1}{2}, \pm\frac{3}{2}$ and $J = \frac{1}{2},\ J_z = \pm\frac{1}{2}$

$$|\frac{3}{2}, \pm\frac{3}{2} > = \mp\, |1, \pm 1 > |\frac{1}{2}, \pm\frac{1}{2} >$$

$$|\frac{3}{2}, \pm\frac{1}{2} > = \mp\, \frac{1}{\sqrt{3}} \left(|1, \pm 1 > |\frac{1}{2}, \mp\frac{1}{2} > \mp\sqrt{2}|1, 0 > |\frac{1}{2}, \pm\frac{1}{2} > \right)$$

$$|\frac{1}{2}, \pm\frac{1}{2} > = \mp\, \frac{1}{\sqrt{3}} \left(\sqrt{2}|1, \pm 1 > |\frac{1}{2}, \mp\frac{1}{2} > \pm|1, 0 > |\frac{1}{2}, \pm\frac{1}{2} > \right).$$

The phenomenological effective mass Luttinger Hamiltonian for the valence bands used in applications for crystals having the cubic symmetry

$$H = -\frac{\hbar^2}{2m_0} \left(\mathbf{p}^2 + m_1 (\mathbf{pJ})^2 + \Delta J(J+1) - \frac{5}{4} \right)$$

describes three double degenerate valence bands around $\mathbf{k} = 0$. In Fig. 2.18 such a split valence band structure is shown schematically.

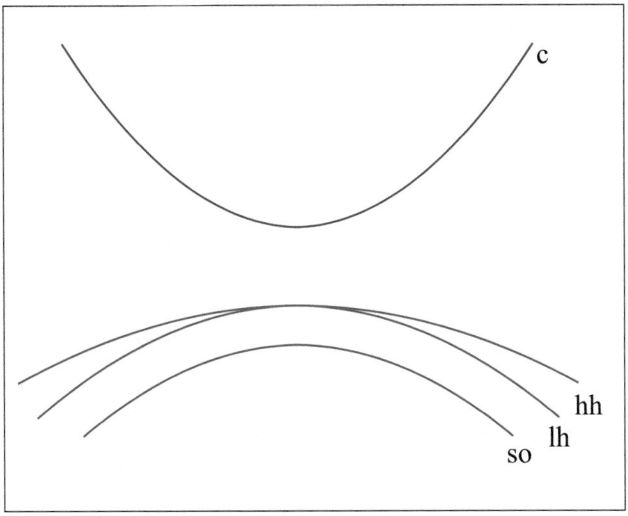

Fig. 2.18 Split valence bands

2.5 Electron States Due to Deviations from Periodicity

2.5.1 *Effective Mass Approximation*

Let us consider a deviation from the crystalline periodicity due to a non-periodical potential $V(\mathbf{r})$ that varies slowly on the scale of the lattice constants. It may be due to a defect in the lattice or a foreign atom. Then we may approximate

$$< n, \mathbf{R}|V|n', \mathbf{R}' >= \int d\mathbf{x} w_{n\mathbf{R}}(\mathbf{r})^* V(\mathbf{r}) w_{n'\mathbf{R}'}(\mathbf{r}) \approx V(\mathbf{R})\delta_{n,n'}\delta_{\mathbf{R},\mathbf{R}'} .$$

We may look for the eigenstates in the presence of both the periodical potential $U(\mathbf{r})$ and this potential $V(\mathbf{r})$ in terms of Wannier functions

$$\frac{1}{\sqrt{\mathcal{N}}} \sum_{\mathbf{R}} \chi(\mathbf{R}) w_{n\mathbf{R}}(\mathbf{r}) .$$

The eigenvalue equation then looks as

$$\sum_{\mathbf{R}'} \left(\tilde{E}(\mathbf{R} - \mathbf{R}') + (V(\mathbf{R}) - \varepsilon)\delta_{\mathbf{R},\mathbf{R}'} \right) \chi(\mathbf{R}') = 0$$

with

$$\tilde{E}_n(\mathbf{R}) \equiv t_n(\mathbf{R}) = \frac{1}{\mathcal{N}} \sum_{\mathbf{k}} E_n(\mathbf{k}) e^{\imath \mathbf{k}\mathbf{R}} .$$

On the other hand, if we introduce the interpolating function

$$\chi(\mathbf{r}) \equiv \sum_{\mathbf{k} \in BZ} e^{\imath \mathbf{k}\mathbf{r}} \tilde{\chi}(\mathbf{k})$$

for all $\mathbf{r}$, then

$$\sum_{\mathbf{R}'} \tilde{E}(\mathbf{R} - \mathbf{R}')\chi(\mathbf{R}') = \sum_{\mathbf{k}} E_n(\mathbf{k})\tilde{\chi}(\mathbf{k}) e^{\imath \mathbf{k}\mathbf{R}}$$
$$= \sum_{\mathbf{k}} \tilde{\chi}(\mathbf{k}) E_n(-\imath\nabla) e^{\imath \mathbf{k}\mathbf{r}}|_{\mathbf{r}=\mathbf{R}} = E_n(-\imath\nabla)\chi(\mathbf{r})|_{\mathbf{r}=\mathbf{R}} .$$

Assuming that $\tilde{\chi}(\mathbf{k})$ vanishes rapidly far away from the band extremum at $\mathbf{k}_0$, where the expansion

$$E_n(\mathbf{k}) \approx E_n(\mathbf{k}_0) + \frac{\hbar^2}{2m^*}(\mathbf{k} - \mathbf{k}_0)^2 + \dots$$

is valid and defining $\chi(\mathbf{r}) = e^{\imath \mathbf{k}_0 \mathbf{r}} \varphi(\mathbf{r})$ we get the Schrödinger equation with the effective mass m^*

$$\left\{ E_n(\mathbf{k}_0) - \frac{\hbar^2}{2m^*} \nabla^2 + V(\mathbf{r}) - \varepsilon \right\} \varphi(\mathbf{r}) = 0 .$$

Under the above assumptions we have also

$$\frac{1}{\sqrt{\mathcal{N}}} \sum_{\mathbf{R}} \chi(\mathbf{R}) w_{n\mathbf{R}}(\mathbf{r}) \approx \varphi(\mathbf{r}) \frac{1}{\sqrt{\mathcal{N}}} \sum_{\mathbf{R}} e^{\imath \mathbf{k}_0 \mathbf{R}} w_{n\mathbf{R}}(\mathbf{r}) = \varphi(\mathbf{r}) u_{n,\mathbf{k}_0}(\mathbf{r}) .$$

2.5.2 *Intrinsic Semiconductors at Finite Temperatures*

Since the importance of semiconductors in modern technological applications is overwhelming we devote more space to their discussion. Pure semiconductor materials, without impurities are called intrinsic. Here the last two relevant bands (valence and conduction bands) are neither completely filled nor empty at ordinary temperatures. The electronic occupation of the Bloch states with wave vector **k** and spin σ in equilibrium at inverse temperature β is described by the Fermi distribution

$$f_{n,\mathbf{k},\sigma} \equiv \langle n_{n,\mathbf{k},\sigma} \rangle = \frac{1}{e^{\beta(\varepsilon_{n,\mathbf{k},\sigma} - \mu)} + 1} .$$

It is meaningful to define the conduction band electrons as "true" electrons

$$f^e_{\mathbf{k},\sigma} \equiv f_{c,\mathbf{k},\sigma} = \frac{1}{e^{\beta(\varepsilon_{c,\mathbf{k},\sigma} - \mu)} + 1} ,$$

while in the valence band it is useful to define the population by the holes (absence of valence band electrons). With these definitions at $T = 0$ we have no holes and no electrons, i.e., an electron–hole vacuum. As we shall see later, this terminology is extremely useful.

At the band edges of a direct gap semiconductor we have in the effective mass approximation the electron energy (in the conduction band) as

$$\varepsilon^e_{\mathbf{k}} = \varepsilon_{c,\mathbf{k},\sigma} = \frac{1}{2} E_g + \frac{\hbar^2}{2m^*_e} k^2 \qquad (m^*_e > 0),$$

while we may define the energy of a hole (in the valence band) as

$$\varepsilon^h_{\mathbf{k}} = -\varepsilon_{v,-\mathbf{k},-\sigma} = \frac{1}{2} E_g + \frac{\hbar^2}{2m^*_h} k^2 \qquad (m^*_h = -m^*_v > 0).$$

This last definition corresponds to the fact that removing an electron of charge $e < 0$, spin σ, wave vector $\mathbf{k}$, and energy $\varepsilon_{v,\mathbf{k},\sigma}$ is equivalent to adding a particle with charge $-e > 0$, spin $-\sigma$, wave vector $-\mathbf{k}$ and energy $-\varepsilon_{v,-\mathbf{k},-\sigma}$. Thus, holes may be looked at as positively charged particles with positive effective mass. In a homogeneous electric field, they would be accelerated accordingly.
The average number of conduction electrons and holes is the same, i.e., the system is electrically neutral

$$\sum_{\mathbf{k},\sigma} f^e_{\mathbf{k},\sigma} = \sum_{\mathbf{k},\sigma} f^h_{\mathbf{k},\sigma}.$$

If the electron–hole occupation of states far away of the band extrema is negligible, one may extrapolate the effective mass approximation overall and we have in the thermodynamic limit

$$\frac{1}{(2\pi)^3}\int d\mathbf{k}\left[\frac{1}{e^{\beta(\frac{\hbar^2k^2}{2m_e}+\frac{1}{2}E_g-\mu)}+1}-\frac{1}{e^{\beta(\frac{\hbar^2k^2}{2m_h}+\frac{1}{2}E_g+\mu)}+1}\right]=0.$$

Since the band gap is still much bigger than the thermal energy ($\beta E_g \ll 1$) one may approximate the Fermi distribution through its non-degenerate limit and therefore

$$\frac{1}{(2\pi)^3}\int d\mathbf{k}\left[e^{-\beta(\frac{\hbar^2k^2}{2m_e}+\frac{1}{2}E_g-\mu)}-e^{-\beta(\frac{\hbar^2k^2}{2m_h}+\frac{1}{2}E_g+\mu)}\right]=0.$$

These integrals may be solved exactly and give rise to a chemical potential sitting in the gap at $\mu = \frac{3}{4}k_BT\ln\left(\frac{m_h}{m_e}\right)$ and the average number of each of the newly defined particles (carrier density) is

$$\langle n_0\rangle = \sqrt{2\pi}(m_em_h)^{\frac{3}{4}}(\beta\hbar^2)^{-\frac{3}{2}}e^{-\frac{1}{2}\beta E_g}.$$

2.5.3 Ionic Impurities

Let us see now what kind of modifications are introduced in the spectrum by putting an extra positive or negative ion in the crystal. Obviously, a minor modification of the continuous bands is not relevant. However, extra discrete states in the forbidden gap may be very important. Therefore, vicinities of band extrema must be looked up. Let us consider a positive ion in the origin. The stationary Schrödinger equation for an electron in the effective mass approximation in the neighborhood of a conduction band minimum is

$$\left\{ E_c - \frac{\hbar^2}{2m^*}\nabla^2 - \frac{e^2}{\epsilon|\mathbf{r}|} - E \right\} \varphi(\mathbf{r}) = 0 \qquad (m^* > 0),$$

where ϵ is the dielectric constant of the crystal and E_c denotes the energetic position of the band minimum. The ground state energy of this hydrogen-like Coulomb problem is well known to be given by the Rydberg energy E_R, here modified by the replacement of the electron mass by his effective mass m^* and the presence of the dielectric constant ϵ of the semiconductor. The binding energy here is the distance of the discrete level from the conduction band

$$E - Ec = -E_R \qquad \left(E_R = \frac{e^2}{2\epsilon a_B}; \quad a_B = \frac{\epsilon\hbar^2}{m^* e^2} \right).$$

The energy of this localized state therefore lies below the band minimum, also in the gap.

In the case of a negative ion placed at the origin we may expect analogously a new state in the band gap near a band maximum described by the equation

$$\left\{ E_V - \frac{\hbar^2}{2m^*}\nabla^2 + \frac{e^2}{\epsilon|\mathbf{r}|} - E \right\} \varphi(\mathbf{r}) = 0 \qquad (m^* < 0)$$

with a negative effective mass m^*. This may be rewritten as

$$\left\{ -E_V - \frac{\hbar^2}{2|m|^*}\nabla^2 - \frac{e^2}{\epsilon|\mathbf{r}|} + E \right\} \varphi(\mathbf{r}) = 0$$

giving rise to the ground state energy

$$E - E_V = E_R$$

lying in the gap above the band maximum. These states in the forbidden energy zone as they are illustrated in Fig. 2.19 are extremely important in the case of semiconductors.

In most of the semi-conducting crystals the dielectric constant is much greater than unity and the effective mass is much smaller than the true electron mass and, as a consequence, the Bohr radius a_B is much bigger than the lattice constant. Thus, the assumption about the smoothness of the potential, as well as the use of the macroscopic dielectric constant ϵ of the material are justified. On the other hand, in a semiconductor the Rydberg energy is much smaller than the energy gap and of course even much smaller than the hydrogenic Rydberg energy.

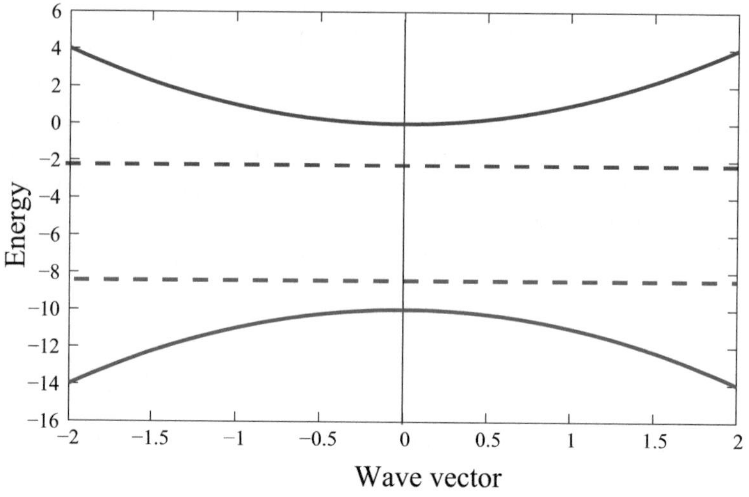

Fig. 2.19 Localized states in the energy gap of a semiconductor

2.5.4 *Extrinsic Semiconductors at Finite Temperatures: Acceptors and Donors*

As we already discussed, foreign ions in a semiconductor crystal lattice give rise to localized states in the band gap. On the other hand, one includes not ions, but neutral atoms. The question is, what happens if the atom gets ionized and losses or gains an electron in favor of the collectivized band electrons? A foreign atom may create a localized state just below the conduction band and remain neutral in the ground state. This would be interpreted as a positive ion state occupied by an electron. Another possibility is to create a localized state just above the valence band. Then in the ground state it will lose its electron in favor of the lower lying valence band continuum. This would be interpreted as a positive unoccupied ion. In the first case we speak about donors, while in the second about acceptors.

Bloch states with a given wave vector $\mathbf{k}$ may be occupied by two electrons of opposite spin. Double occupation of localized states, however, may be forbidden. Indeed, the average Coulomb repulsion energy for two electrons on the same site is

$$U \equiv \frac{e^2}{\epsilon} \int d\mathbf{r} \int d\mathbf{r}' \frac{|\psi_\uparrow(\mathbf{r})|^2 |\psi_\downarrow(\mathbf{r}')|^2}{|\mathbf{r} - \mathbf{r}'|}.$$

In the case of the Coulomb ground state ($l = 0, \quad n = 1$)

$$\psi_\uparrow(\mathbf{r}) = \psi_\downarrow(\mathbf{r}) = \frac{1}{\sqrt{\pi a_B^3}} e^{-\frac{r}{a_B}}$$

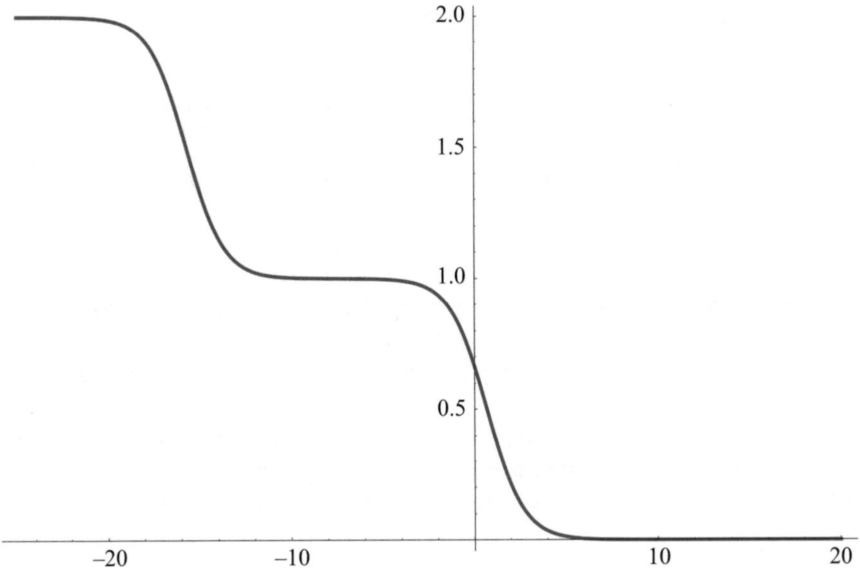

Fig. 2.20 The electron occupation $\mathscr{F}_D$ of a donor state

the integral may be solved analytically giving rise to

$$U = \frac{5}{8}\frac{e^2}{\epsilon a_B} = \frac{5}{4}E_R.$$

Therefore, the energy of the double occupied state may differ essentially from twice the energy of the single occupation. In this case one may write the energy of the donor state as in terms of its occupation by electrons as

$$E_D(n_\uparrow, n_\downarrow) = \varepsilon_D(n_\uparrow + n_\downarrow) + Un_\uparrow n_\downarrow; \qquad (n_\uparrow, n_\downarrow = 0, 1).$$

The average occupation at a finite temperature is given by

$$\langle n_\uparrow \rangle \equiv \langle n_\downarrow \rangle = \frac{\sum_{n_\uparrow=0,1}\sum_{n_\downarrow=0,1} n_\uparrow e^{-\beta[E_D(n_\uparrow,n_\downarrow)-\mu(n_\uparrow+n_\downarrow)]}}{\sum_{n_\uparrow=0,1}\sum_{n_\downarrow=0,1} e^{-\beta[E_D(n_\uparrow,n_\downarrow)-\mu(n_\uparrow+n_\downarrow)]}}$$

and

$$\langle n \rangle \equiv \langle n_\uparrow \rangle + \langle n_\downarrow \rangle = 2\frac{e^{-\beta(\varepsilon_D-\mu)} + e^{-\beta[2(\varepsilon_D-\mu)+U]}}{1 + 2e^{-\beta(\varepsilon_D-\mu)} + e^{-\beta[2(\varepsilon_D-\mu)+U]}} \equiv \mathscr{F}_D$$

instead twice of the Fermi function one would have without repulsion. This new distribution is illustrated in Fig. 2.20.

In the extreme case of very low temperatures $\frac{U}{k_B T} \to \infty$ it gives

$$\langle n \rangle = 2 \frac{e^{-\beta(\varepsilon_D - \mu)}}{1 + 2e^{-\beta(\varepsilon_D - \mu)}} = \frac{1}{e^{\beta(\varepsilon_D - \mu - k_B T \ln 2)} + 1},$$

which looks like a Fermi function with a shifted chemical potential

$$\mu \to \mu + k_B T \ln 2.$$

Analogously we have for the positively charged holes on an acceptor level

$$E_A(m_\uparrow, m_\downarrow) = \varepsilon_A(m_\uparrow + m_\downarrow) + U m_\uparrow m_\downarrow; \qquad (m_\uparrow, m_\downarrow = 0, 1)$$

with $m_\uparrow$, $m_\downarrow$ being the hole occupations. One has still to take into account that the chemical potential for the holes is $-\mu$ and $\varepsilon_A = \frac{1}{2} E_g - E_R$.
The equation for the electro-neutrality in an extrinsic semiconductor looks therefore as

$$2 \sum_{\mathbf{k}} f_{\mathbf{k}}^e + \sum_D \mathscr{F}_D - N_D = 2 \sum_{\mathbf{k}} f_{\mathbf{k}}^h + \sum_A \mathscr{F}_A - N_A.$$

This equation may be solved for the chemical potential only numerically. As illustration, we show on Fig. 2.21 the one-electron state density and its occupation in an n-type semiconductor with more donors (D) as acceptors (A) at $T = 0$. The green filling shows the (simple or double) occupation of the states. The donor states lying higher in energy lose some of their electrons in favor of the acceptor sates.

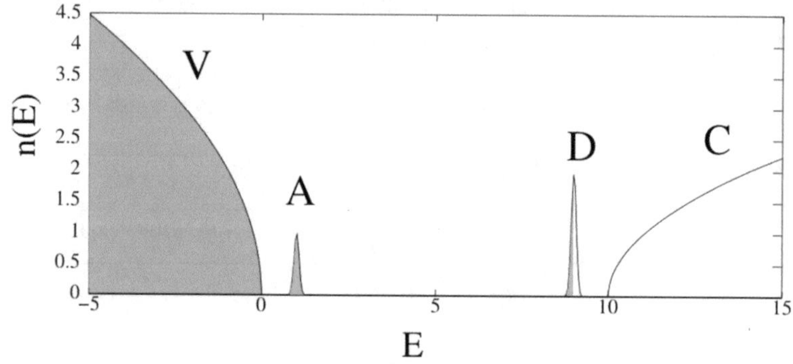

Fig. 2.21 State density in a compensated n-type semiconductor at $T = 0$

2.6 Semiconductor Contacts

Until now we discussed only bulk properties of solids; however, phenomena at the interfaces play an overwhelming role in all solid-state devices. We describe in the following two such simple cases: the penetration of a static electric field in a semiconductor and the contact between an n-type extrinsic semiconductor and a p-type extrinsic semiconductor of the same crystal. In both cases thermal equilibrium is assumed.

A consequent quantum mechanical discussion of these problems would be extremely difficult. A finite semiconductor is not a periodical crystal and the eigenstates and eigenenergies differ from that of the infinite one. A simplified quasi-classical approach, however, has proved to be very successful. We need to describe a situation in which the quantum mechanical nature of the crystals, as well as macroscopic inhomogeneities must be reconciled. The basic object here to consider is the Wigner function depending on the momentum $\mathbf{p}$ and coordinate $\mathbf{x}$. It is defined as

$$f(\mathbf{p}, \mathbf{x}, t) = \int d\mathbf{y} e^{\frac{\imath \mathbf{p}\mathbf{y}}{\hbar}} \langle \psi(\mathbf{x} + \frac{1}{2}\mathbf{y}, t)^{+} \psi(\mathbf{x} - \frac{1}{2}\mathbf{y}, t) \rangle,$$

where $\psi(\mathbf{x}, t)$ is the second quantized wave function and $\langle \ldots \rangle$ means averaging over a given ensemble. This implies the normalization

$$\int d\mathbf{x} \int \frac{d\mathbf{p}}{(2\pi\hbar)^3} f(\mathbf{p}, \mathbf{x}, t) = \langle N \rangle,$$

where N is the operator of the total number of particles. It may be shown that all the averages of operators which are a sum of two operators $O(\mathbf{x}, \frac{\hbar}{\imath}\nabla) \equiv A(\mathbf{x}) + B(\frac{\hbar}{\imath}\nabla)$ depending on the coordinate, respectively, on the momentum may be expressed as the integrals

$$\langle O(\mathbf{x}, \frac{\hbar}{\imath}\nabla) \rangle = \int d\mathbf{x} \int \frac{d\mathbf{p}}{(2\pi\hbar)^3} f(\mathbf{p}, \mathbf{x}, t) O(\mathbf{x}, \mathbf{p}).$$

However, generally speaking $f(\mathbf{p}, \mathbf{x})$ is neither real, nor positive. Nevertheless, in the quasi-classical limit ($\hbar \to 0$) it may be shown that it will be real and positive. The Hamiltonian of a particle in the presence of a potential $U(\mathbf{x})$

$$H = -\frac{\hbar^2 \nabla^2}{2m} + U(\mathbf{x})$$

has the above described structure.

One assumes tacitly that the corresponding Wigner function in this quasi-classical limit, in thermal equilibrium is given again by the Fermi distribution

$$f(\mathbf{p}, \mathbf{x}) = \frac{1}{e^{\beta\left(\frac{p^2}{2m}+U(\mathbf{x})-\mu\right)} + 1}$$

in the phase space (**x**, **p**).

2.6.1 Electric Field Penetration into a Semiconductor

In a previous discussion we have admitted implicitly that a homogeneous constant electric field may be created in an infinite crystal. This is an extreme idealization. It is well known that metals screen out static electric fields, while in a semiconductor, it may still penetrate at a finite macroscopic depth. We shall discuss here this last case within the quasi-classical approach. Let us consider the surface of an intrinsic semiconductor characterized by a dielectric constant ϵ and an equal number N of electrons and holes in the macroscopic volume Ω. Their density in equilibrium n_0 is determined by the band gap E_g and the temperature T and for $E_g \gg k_B T$ it is $n_0 \equiv N/\Omega \sim e^{-\frac{E_g}{2k_B T}}$.
Let us consider that the semiconductor occupies the half space $x > 0$. The static electric potential V(x) satisfies he Poisson equation outside

$$\nabla^2 V(\mathbf{x}) = 0 \quad (x < 0)$$

respectively, inside the semiconductor

$$\epsilon \nabla^2 V(\mathbf{x}) = -4\pi\rho(\mathbf{x}) \quad (x > 0),$$

where $\rho(\mathbf{r})$ is the charge density of carriers (electrons and holes). In the presence of an external electric field $\mathscr{E}$ the potential outside the semiconductor is

$$V(\mathbf{x}) = -\mathscr{E}x + const. \qquad (x < 0) \; .$$

While in the absence of the field the charge density vanishes, in the presence of the field in equilibrium at a finite temperature it is determined self-consistently by the potential itself. In a quasi-classical approach (at high temperatures) the Maxwell–Boltzmann distribution gives the probability density of a state **r**, **v** in the phase space

$$f(\mathbf{v}, \mathbf{x}) = \frac{e^{-\beta(\frac{mv^2}{2}+eV(\mathbf{x}))}}{\int d\mathbf{v}' \int d\mathbf{x}' e^{-\beta(\frac{mv'^2}{2}+eV(\mathbf{x}'))}} .$$

This defines the equilibrium charge density of electrons and holes as

$$\rho_{e,h}(\mathbf{x}) = \pm eN \frac{e^{\mp \frac{eV(\mathbf{x})}{k_B T}}}{\int d\mathbf{r}' e^{\mp \frac{eV(\mathbf{x}')}{k_B T}}}.$$

Under the assumption that the potential vanishes deep in the semiconductor, the integral in the denominator is proportional to the volume Ω and for a macroscopic volume we get approximately

$$\rho_{e,h}(\mathbf{x}) = \pm e n_0 e^{\mp \frac{eV(\mathbf{x})}{k_B T}}$$

and for $eV \ll k_B T$ we have the charge density

$$\rho(\mathbf{x}) = \rho_e(\mathbf{x}) + \rho_h(\mathbf{x}) = -2en_0 \frac{eV(\mathbf{x})}{k_B T}.$$

Therefore, we get the self-consistent equation

$$\epsilon \nabla^2 V(\mathbf{x}) = \frac{8\pi e^2 n_0}{k_B T} V(\mathbf{x})$$

for the determination of the potential inside the semiconductor. By introducing the Debye length

$$L_D \equiv \sqrt{\frac{k_B T \epsilon}{8\pi e^2 n_0}}$$

the equation looks as

$$\nabla^2 V(\mathbf{x}) - \frac{1}{L_D^2} V(\mathbf{x}) = 0.$$

Now we must still consider the electromagnetic boundary condition

$$\frac{\partial V(x)}{\partial x}|_{x=-0} = \epsilon \frac{\partial V(x)}{\partial x}|_{x=+0}.$$

to connect the solutions inside and outside. This gives rise to

$$V(x) = \mathscr{E} \frac{L_D}{\epsilon} e^{-\frac{x}{L_D}}$$

for $x > 0$ and it is illustrated in Fig. 2.22.

To conclude: Due to the screening by the carriers, the field penetrates at a depth L_D into the semiconductor. At room temperatures, this Debye length may be of order of centimeters and therefore our macroscopic treatment, as well as the

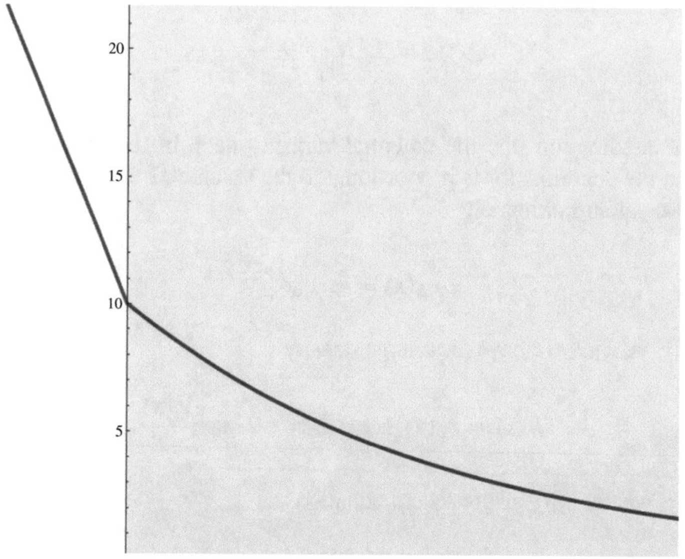

Fig. 2.22 Penetration of the potential into the semiconductor

existence of a more or less homogeneous electric field inside a semiconductor gets justified.

2.6.2 *p–n Contact*

Let us now consider the contact of two semiconductors of the same material, but one of them is p-doped (i.e., with acceptors), while the other one is n-doped (i.e., with donors). Again, we consider the contact surface to be a plane transverse to the x-axis at $x = 0$. On the left side is the p-type semiconductor, while on the right side the n-type.

Since we want to discuss here both non-degenerate as well as degenerate cases, we must use the quasi-classical Fermi distribution for fermions defined above.

So long the two semiconductors are kept apart, we have for the bulk materials a homogeneous situation with the electro-neutrality condition for the electrons, holes in the bands and the negative, respectively, positive ionized acceptors and donors.

$$\int \frac{d\mathbf{p}}{(2\pi\hbar)^3}\left(f_e(\mathbf{p}) - f_h(\mathbf{p})\right) + nf_-^A = 0$$

in the p-material and

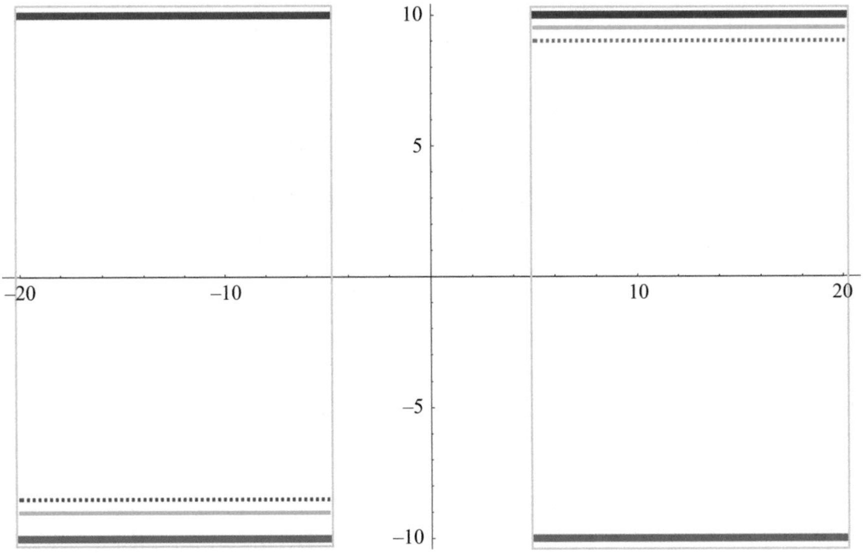

Fig. 2.23 Separated p and n semiconductors at $T = 0$

$$\int \frac{d\mathbf{p}}{(2\pi\hbar)^3} \left(f_e(\mathbf{p}) - f_h(\mathbf{p})\right) - n_D f_+^D = 0$$

in the n-material. (Here f_e, f_h and f_-^A, f_+^D are the Fermi distributions of electrons and holes in the bands, respectively, on the donors and acceptors.) The chemical potentials in these two equations μ_p and μ_n are of course different and $\mu_p < \mu_n$.

Now let us consider the simplest case of zero temperature (see Fig. 2.23). In the ground state of the p-type semiconductor the valence band (red) is full occupied, while the acceptor level (dotted blue line) and the conduction band (edge—blue line) are empty, the chemical potential (green line) lies somewhere between the valence band and the acceptor level. In the ground state of the n-type semiconductor the valence band (edge—red line) and the donor level (dotted red line) are occupied and the chemical potential (green line) lies between the donor level and the conduction band.

When the contact between the semiconductors is established in thermal equilibrium at T $= 0$ one has a single system where the chemical potential μ must be the same overall (green line). This is possible only if at the contact surface a potential jump occurs as it is shown on Fig. 2.24.
Assuming the contact is defined by a plane transverse to the x-axis at $x = 0$, at any finite temperature T one has an in-homogeneous charge density of electrons and holes:

$$\rho(x) = e \int \frac{d\mathbf{p}}{(2\pi\hbar)^3} \cdot \left(f_e(\mathbf{p},x) - f_h(\mathbf{p},x)\right) - e n_D \theta(x) f_+^D(x) + e n_A \theta(-x) f_-^A(x)$$

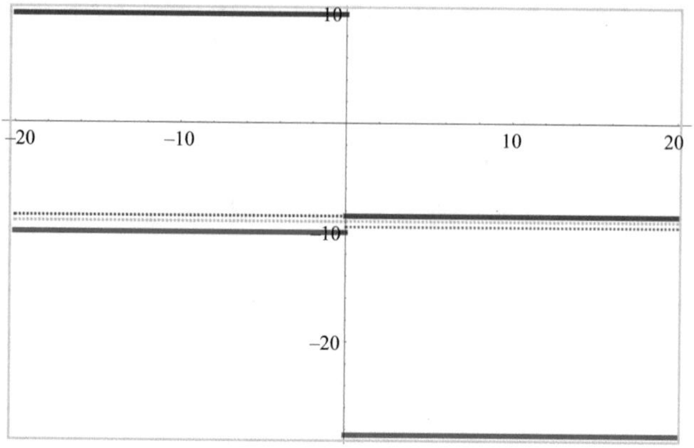

Fig. 2.24 p and n semiconductors in contact at $T = 0$

with

$$f_e(\mathbf{p}, x) = \frac{1}{e^{\beta\left(\frac{p^2}{2m_e}+eV(x)+\frac{1}{2}Eg-\mu\right)}+1}$$

$$f_h(\mathbf{p}, x) = \frac{1}{e^{\beta\left(\frac{p^2}{2m_h}-eV(x)+\frac{1}{2}Eg+\mu\right)}+1}$$

and

$$f_-^A(x) = \frac{1}{e^{\beta(E_A+eV(x)-\mu)}+1}$$

$$f_+^D(x) = 1 - \frac{1}{e^{\beta(E_D+eV(x)-\mu)}+1} = \frac{1}{e^{\beta(-E_D-eV(x)+\mu)}+1}.$$

Here the reference energy choice was the middle of the gap and therefore E_A and E_D are the distances to the center of the gap.
The potential $V(x)$ has to be determined self-consistently from the Poisson equation

$$\epsilon \frac{d^2}{dx^2} V(x) = -4\pi\rho(x)$$

with the condition of continuity of the derivative of the potential at $x = \pm 0$:

$$\frac{d}{dx} V(x)|_{x=-0} = \frac{d}{dx} V(x)|_{x=+0}$$

(since the dielectric constants of the two materials were assumed to be the same). The potential energy has to be assumed also to be continuous and therefore $V(-0) = V(+0)$. To solve the Poisson equation (a second order differential equation) one still needs two initial conditions. Usually one gives the initial value of the function and its derivative in the initial point. Here the peculiarity of the physics defines two different conditions to impose, namely the values at $\pm\infty$, that should correspond to the situation in the separated semiconductors. These two conditions define uniquely the solution; however, these equations are not solvable analytically and the numerical method to solve it must be correspondingly adapted.

We have seen before the solution for $T = 0$. For higher temperatures (in the physically relevant situations this is just the room temperature) one may consider the non-degenerate case that simplifies essentially the equations.

Indeed in the non-degenerate case the charge density is given by

$$\rho(x) = e \int \frac{d\mathbf{p}}{(2\pi\hbar)^3} \left(e^{-\beta\left(\frac{p^2}{2m_e} + eV(x) + \frac{1}{2}Eg - \mu\right)} - e^{-\beta\left(\frac{p^2}{2m_h} - eV(x) + \frac{1}{2}Eg + \mu\right)} \right)$$

$$+ \theta(x) e n_D e^{\beta(E_D + eV(x) - \mu)} - \theta(-x) e n_A e^{-\beta(E_A + eV(x) - \mu)}.$$

The integrals over the momenta may be performed and we get

$$\begin{aligned}\rho(x) &= e n_{0e} e^{-\beta(eV(x)-\mu)} - e n_{0h} e^{\beta(eV(x)-\mu)} \\ &\quad - \theta(x) e n_D^0 e^{\beta(eV(x)-\mu)} + \theta(-x) e n_A^0 e^{-\beta(eV(x)-\mu)},\end{aligned}$$

where $n_{0e} = \left(\frac{2\pi\hbar^2}{k_B T m_e}\right)^{\frac{3}{2}} e^{-\beta\frac{1}{2}E_g}$, $n_{0h} = \left(\frac{2\pi\hbar^2}{k_B T m_e}\right)^{\frac{3}{2}} e^{-\beta\frac{1}{2}E_g}$, and $n_D^0 = n_D e^{\beta E_D}$, $n_A^0 = n_A e^{-\beta E_A}$. (Actually, only the non-degeneracy of the distribution in the bands is essential for this step!)

We may require a total charge neutrality on both sides (no macroscopic polarization!). Taking into account that the potential differs from constants $V(\pm\infty)$ only in a finite domain in the vicinity of the contact, we may write

$$\begin{aligned} e\left(n_{0e} e^{-\beta(eV(-\infty)-\mu)} - e n_{0h} e^{\beta(eV(-\infty)-\mu)} + n_A^0 e^{-\beta(eV(-\infty)-\mu)}\right) &= 0 \\ e\left(n_{0e} e^{-\beta(eV(+\infty)-\mu)} - e n e_h^{\beta(eV(+\infty)-\mu)} - n_D^0 e^{-\beta(eV(+\infty)-\mu)}\right) &= 0. \end{aligned}$$

So long the semiconductors were separated both equations were satisfied with $V(\pm\infty) = 0$ and the chemical potentials μ_p, respectively, μ_n of the two materials, respectively.

After the contact has been established one may still choose $V(-\infty) = 0$ and therefore, since far away from the contact the position of the chemical potential relative to the bands has to be the same as before, one must take $\mu \equiv \mu_n$. On the other hand, the same must be true at $x \to \infty$ for the other semiconductor. It follows that simultaneously $\mu \equiv \mu_n + eV(\infty)$. Therefore $eV(\infty) = \mu_p - \mu_n$. Then, not

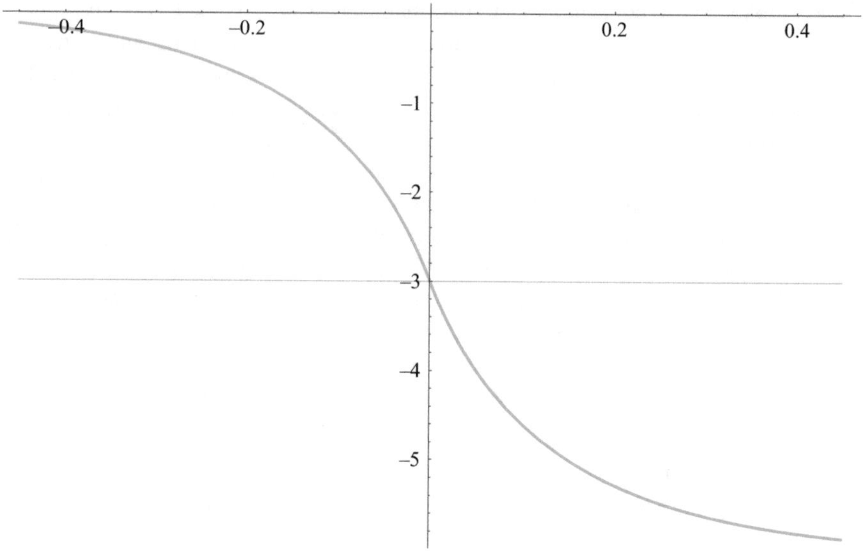

Fig. 2.25 $\beta eV(x)$ in the non-degenerate case (with $c = 3$)

only the electro-neutrality is overall satisfied, but also the charge density vanishes at $x \to \pm\infty$ and all the parameters are well defined.

The non-degenerate solution obtained numerically for the simplified symmetrical case of $n_A = n_D$ shows on a convenient length scale a smoothing of the potential drop at the contact as in Fig. 2.25 with the choice $c \equiv \frac{1}{2}\beta(\mu_n - \mu_p) = 3$.

To conclude, the p–n contact in equilibrium creates a potential barrier for the carriers. An application of an external electric potential either lowers or increases this barrier and therefore the current flow depends on the applied polarity. The p–n contact acts as a rectifier.

Chapter 3
Electron–Electron Interaction

The potential a valence electron feels in a crystal is not a predetermined one. The electrons interact with each other by Coulomb forces and the one-electron theory is to be understood at most in the sense of a self-consistent approximation. Diagrammatic perturbation theory may lead to higher many-body corrections. In a semiconductor an electron in the conduction band may move "freely," while in the valence band the same is true for the holes (absent electrons). This leads to the picture of an electron–hole plasma. A single electron–hole pair by Coulomb attraction may be bound to an exciton with hydrogen-like spectrum.

3.1 The Exciton

3.1.1 Wannier Exciton

Until now we discussed the electron states and their occupation in a periodical lattice under the assumption of no interaction between the electrons. We nevertheless made an exception to correct the double occupation of the localized states. In this chapter, we shall discuss other many-electron aspects. The next simplest problem is the Coulomb attraction between a conduction band electron and a valence band hole in a semiconductor. Obviously these two may form a bound state, called exciton. As in the description of the impurity states we use the effective mass approximation in the neighborhood of the band extrema. By separating the center of mass motion

$$\Psi(\mathbf{r}_e, \mathbf{r}_h) = \phi(\mathbf{r}_{CM})\chi(\mathbf{r}_{rel})$$

L. A. Bányai, *A Compendium of Solid State Theory*,
https://doi.org/10.1007/978-3-030-37359-7_3

$$\mathbf{r}_{CM} \equiv \frac{m_e^* \mathbf{r}_e + m_h^* \mathbf{r}_h}{m_e^* + m_h^*}; \qquad \mathbf{r}_{rel} \equiv \mathbf{r}_e - \mathbf{r}_h$$

$$\phi(\mathbf{r}_{CM}) = \frac{1}{\sqrt{V}} e^{\imath \mathbf{k}_{CM} \mathbf{r}_{CM}} \quad ,$$

the relative motion is described by the Wannier equation

$$\left\{ -\frac{\hbar^2}{2\mu} \nabla^2 - \frac{e^2}{\epsilon |\mathbf{r}|} + E_g - E \right\} \chi(\mathbf{r}) = 0,$$

where the reduced mass is given by $\frac{1}{\mu} = \frac{1}{m_e^*} + \frac{1}{m_h^*}$. As it is well known from the hydrogen atom problem, the eigenstates are characterized by the quantum numbers $n = 1, 2, \ldots, \quad l = 0, 1 \ldots, \quad m = 0 \pm 1, \ldots$, while the eigenenergies depend only on the main quantum number n. The ground state wave function is

$$\psi_{1,0} = \frac{1}{\sqrt{\pi} a_B^{\frac{3}{2}}} e^{-\frac{r}{a_B}}$$

with the ground state energy

$$E_n^x = E_g - E_R \ ,$$

where the Bohr radius and the Rydberg energy are

$$a_B = \frac{\hbar^2 \epsilon}{e^2 \mu}, \quad E_R = \frac{e^2}{2\epsilon a_B}.$$

However, as we already discussed by the ionic impurity states, the scales of energy and radius are essentially different from that of the hydrogen atom due to the difference in the masses and the dielectric constant.

This state may not be represented in the one-particle state density (it does not belong to the one-electron states), but may be seen in the absorption spectrum. For higher lying states of the pair the above described approximation is, however, not appropriate.

3.1.2 Exciton Beyond the Effective Mass Approximation

We shall describe here an alternative way to treat the exciton states without using the effective mass approximation. The Hamiltonian of the conduction band electrons and valence band holes of spin σ in the second quantization formalism may be defined either through the creation/annihilation operators of the Bloch or of the Wannier states

$$H_0 = \sum_{\sigma=\pm 1} \sum_{\alpha=e,h} \sum_{\mathbf{k}\in BZ} \epsilon_\alpha(\mathbf{k}) a^+_{\alpha,\sigma,\mathbf{k}} a_{\alpha,\sigma,\mathbf{k}}$$
$$= \sum_{\sigma=\pm 1} \sum_{\alpha=e,h} \sum_{\mathbf{R},\mathbf{R}'} t_\alpha(\mathbf{R}-\mathbf{R}') a^+_{\alpha,\sigma,\mathbf{R}} a_{\alpha,\sigma,\mathbf{R}'} ; \qquad (\alpha = e, h) .$$

Here $\epsilon_\alpha(\mathbf{k})$ $(\alpha = e, h)$ are the band energies of electrons respectively, holes. The Hamiltonian of the Coulomb attraction between the electrons and holes, while being defined through the charge densities, is easily expressed in terms of the second quantized wave functions

$$H_{eh} = -\sum_{\sigma,\sigma'} \int d\mathbf{x} \int d\mathbf{x}' \psi_{e,\sigma}(\mathbf{x})^+ \psi_{e,\sigma}(\mathbf{x}) \frac{e^2}{|\mathbf{x}-\mathbf{x}'|} \psi^+_{h,\sigma'}(\mathbf{x}') \psi_{h,\sigma'}(\mathbf{x}') .$$

The second quantized wave functions of conduction band electrons or valence band holes of spin σ may be again defined through the annihilation/creation operators either of Bloch states or Wannier states

$$\psi_{\alpha,\sigma}(\mathbf{x}) = \sum_{\mathbf{k}} e^{\imath \mathbf{k}\mathbf{x}} u_{\alpha\mathbf{k}}(\mathbf{x}) a_{\alpha,\sigma,\mathbf{k}} = \sum_{\mathbf{R}} w_\alpha(\mathbf{x},\mathbf{R}) a_{\alpha,\sigma,\mathbf{R}} ; \quad (\alpha = e, h).$$

In the Wannier version one may use the fact that the Wannier functions are strongly localized on the lattice nodes in order to introduce a simplifying approximation. Namely, to retain in the charge density operator only the terms corresponding to the local densities on the nodes, i.e.,

$$H_{eh} \approx -\sum_{\sigma,\sigma'} \sum_{\mathbf{R},\mathbf{R}'} \int d\mathbf{x} \int d\mathbf{x}' |w_e(\mathbf{x},\mathbf{R})|^2 \frac{e^2}{|\mathbf{x}-\mathbf{x}'|} |w_h(\mathbf{x},\mathbf{R}')|^2 a^+_{e,\sigma,\mathbf{R}} a_{e,\sigma,\mathbf{R}} a^+_{h,\sigma,\mathbf{R}'} a_{h,\sigma,\mathbf{R}'} .$$

Furthermore, since the Coulomb potential varies slowly on the scale of the elementary cell we may even write

$$H_{eh} \approx -\sum_{\sigma=\pm 1} \sum_{\sigma'=\pm 1} \sum_{\mathbf{R}\neq\mathbf{R}'} \frac{e^2}{|\mathbf{R}-\mathbf{R}'|} a^+_{e,\sigma,\mathbf{R}} a_{e,\sigma,\mathbf{R}'} a^+_{h,\sigma',\mathbf{R}'} a_{h,\sigma',\mathbf{R}}$$
$$- \sum_{\sigma=\pm 1} \sum_{\sigma'=\pm 1} c_{eh} \sum_{\mathbf{R}} a^+_{e,\sigma,\mathbf{R}} a_{e,\sigma,\mathbf{R}} a^+_{h,\sigma',\mathbf{R}} a_{h,\sigma',\mathbf{R}} .$$

Here we have separated the on-site terms $\mathbf{R} = \mathbf{R}'$ since in that case one has still to compute the finite integral

$$c_{eh} = \int d\mathbf{x} \int d\mathbf{x}' |w_e(\mathbf{x}, \mathbf{R})|^2 \frac{e^2}{|\mathbf{x} - \mathbf{x}'|} |w_h(\mathbf{x}, \mathbf{R})|^2 .$$

One may encounter two extreme cases:

(i) either the on-site term c_{eh} dominates, and one may ignore the inter-site terms (Frenkel exciton) with the lattice Hamiltonian

$$H_{Frenkel} = \sum_{\sigma=\pm 1} \sum_{\alpha=e,h} \sum_{\mathbf{R},\mathbf{R}'} t_\alpha(\mathbf{R} - \mathbf{R}') a^+_{\alpha,\sigma,\mathbf{R}} a_{\alpha,\sigma,\mathbf{R}'}$$
$$- \sum_{\sigma=\pm 1} \sum_{\sigma'=\pm 1} c_{eh} \sum_{\mathbf{R}} a^+_{e,\sigma,\mathbf{R}} a_{e,\sigma,\mathbf{R}} a^+_{h,\sigma',\mathbf{R}} a_{h,\sigma',\mathbf{R}} ,$$

(ii) or the inter-site term dominates and we get the Wannier Hamiltonian

$$H_{Wannier} = \sum_{\sigma=\pm 1} \sum_{\alpha=e,h} \sum_{\mathbf{R},\mathbf{R}'} t_\alpha(\mathbf{R} - \mathbf{R}') a^+_{\alpha,\sigma,\mathbf{R}} a_{\alpha,\sigma,\mathbf{R}'}$$
$$- \sum_{\sigma=\pm 1} \sum_{\sigma'=\pm 1} \sum_{\mathbf{R}\neq\mathbf{R}'} \frac{e^2}{|\mathbf{R} - \mathbf{R}'|} a^+_{e,\sigma,\mathbf{R}} a_{e,\sigma,\mathbf{R}'} a^+_{h,\sigma',\mathbf{R}'} a_{h,\sigma',\mathbf{R}} .$$

Accordingly, in the Wannier case we get for the energy of a single electron–hole pair the Hamiltonian matrix

$$h_{\mathbf{R},\mathbf{R}'} = t_e(\mathbf{R} - \mathbf{R}') + t_h(\mathbf{R} - \mathbf{R}') - \frac{e^2}{|\mathbf{R} - \mathbf{R}'|}(1 - \delta_{\mathbf{R},\mathbf{R}'})$$

or after a discrete Fourier transformation (implying that the wave vectors belong to the Brillouin Zone)

$$h_{\mathbf{k},\mathbf{k}'} = (\epsilon_e(\mathbf{k}) + \epsilon_h(\mathbf{k}))\delta_{\mathbf{k},\mathbf{k}'} - V(\mathbf{k} - \mathbf{k}')$$

with

$$V(\mathbf{q}) = \sum_{\mathbf{R}\neq 0} \frac{e^2}{|\mathbf{R}|} e^{\imath \mathbf{q}\mathbf{R}} .$$

What we have gained is that now we included all the band states and therefore, we may compute also, higher lying pair states influenced by the Coulomb attraction. This is illustrated for the Wannier case in Fig. 3.1 showing the calculated exciton state density (electron–hole pair states) with and without Coulomb interaction in a cubic crystal, within the simplified tight-binding model with nearest neighbor coupling (see Sect. 2.3.7). In the absence of Coulomb forces, one may remark in

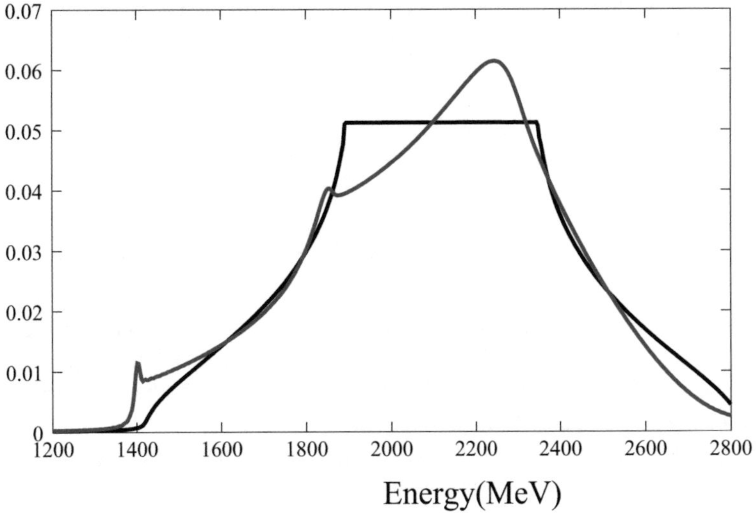

Fig. 3.1 Exciton state density in the whole band calculated within the Wannier model.(black—without Coulomb interaction, blue—with Coulomb interaction)

the middle of the spectrum the sharp flat top (Van Hove singularity), which is a consequence of the discontinuous jumps of the Brillouin zone.

3.2 Many-Body Approach to the Solid State

3.2.1 Self-Consistent Approximations

Until now we have tried to describe the properties of crystals only within the frame of one-particle (or at most two-) particle states in the presence of a periodical potential. This potential, however, may not be identified with the potential created by the lattice of the bare ions, since the electrons at their turn screen the ions. Each electron sees not only the ions, but also the other electrons. To apply the one-particle treatment we need a way to define the periodical potential to be used for the calculation of the band structure.

The most important schemes for the treatment of quantum mechanical many-body systems are the self-consistent approximations we describe here. Within this frame it is possible also to precise the definition of the periodic potential. (In what follows, for the simplicity of notations, we include the coordinate $\mathbf{r}$ as well as the spin $\sigma = \pm\frac{1}{2}$ in the generalized coordinate x, whenever it is meaningful.)

We consider here a system of electrons interacting through repulsive Coulomb forces in the presence of an external potential $U(x)$. The later one includes the attractive Coulomb potential of the positive ions, supposed to stay on a periodical lattice. The dynamics of the ions will be treated later in the chapter about phonons. Within the second quantized formalism (see Chap. 11) the Hamilton operator of this system is

$$H = \int dx \psi(x)^+ \left[-\frac{\hbar^2}{2m}\nabla^2 + U(x) \right] \psi(x)$$
$$+ \frac{1}{2} \int dx \int dx' \psi(x)^+ \psi(x')^+ \frac{e^2}{|x - x'|} \psi(x') \psi(x) \ .$$

The purpose is to reduce the problem in a first approximation to an effective one-particle one (consisting only of products of two creation–annihilation operators) in order to be tractable and close to the previous discussion of the quantum mechanical motion one-particle in a periodical potential.
The simplest scheme consists in "developing" the product of two operators A and B around their averages. Let us consider the identity

$$AB = A\langle B \rangle + \langle A \rangle B - \langle A \rangle \langle B \rangle + (A - \langle A \rangle)(B - \langle B \rangle) \ .$$

Or, ignoring the "fluctuation" around the average values

$$(A - \langle A \rangle)(B - \langle B \rangle)$$

we have

$$AB \approx A\langle B \rangle + \langle A \rangle B - \langle A \rangle \langle B \rangle.$$

Using this approximation for the choice $A \equiv \psi(x)^+ \psi(x)$ and $B \equiv \psi(x')^+ \psi(x')$ in the interaction term we get the one-electron type Hartree Hamiltonian

$$H^{\mathscr{H}} \equiv \int dx \psi(x) \left\{ -\frac{\hbar^2}{2m}\nabla^2 + U^H_{eff}(x) \right\} \psi(x) + const^{\mathscr{H}}$$

with the effective Hartree potential

$$U^{\mathscr{H}}_{eff}(x) \equiv U(x) + \sum_{\sigma'} \int dx' \frac{e^2}{|x - x'|} \langle \psi_{\sigma'}(x')^+ \psi_{\sigma'}(x') \rangle$$

and the Hartree constant

$$const^{\mathscr{H}} = -\frac{1}{2} \int dx \int dx' \frac{e^2}{|x - x'|} \langle \psi_\sigma(x)^+ \psi_\sigma(x) \rangle \langle \psi_{\sigma'}(x')^+ \psi_{\sigma'}(x') \rangle \ .$$

This constant ensures that $\langle H^{\mathscr{H}} \rangle = \langle H \rangle$.

Until now we did not specify the meaning of the averages themselves. They may be defined by their equilibrium values, i.e.,

$$\langle\psi(x)^+\psi(x)\rangle_0 \equiv \frac{1}{Tr\left\{e^{-\beta(H^{\mathscr{H}}-\mu N)}\right\}} Tr\left\{e^{-\beta(H^{\mathscr{H}}-\mu N)}\psi(x)^+\psi(x)\right\}$$

and computed from this definition self-consistently. The result is a temperature dependent effective one-particle spectrum with Hartree eigenfunctions and eigenvalues

$$\left\{-\frac{\hbar^2}{2m}\nabla^2 + U^{\mathscr{H}}_{eff}(x)\right\}\phi(x) = \epsilon\phi(x) \ .$$

The resulting effective Bloch spectrum in the presence of only a periodical potential due to the fixed ions is therefore relevant for a given temperature. The potential to be used in calculations of the Bloch functions and band spectrum in a real crystal is this self-consistent one. One uses mostly the zero temperature (ground state) as reference state.

However, one may tackle also a time-dependent problem within a s.c. approximation, with time-dependent averages $\langle\psi(x)^+\psi(x)\rangle_t$ with initial conditions and additional non-periodical external potentials. In this case the Hartree Hamiltonian is implicitly time dependent. Nevertheless, if the external potential does not depend on time, the average energy is conserved (again due to the special role of the time-dependent Hartree constant!)

$$\frac{\partial}{\partial t}\langle H^{\mathscr{H}}(t)\rangle_t = 0 \,.$$

The choice of the operators A and B in our former reasoning was arbitrary. The proper argument is, that as it may be shown, the Hartree choice gives the best one-particle approximation for the grand canonical distribution conserving the total number of fermions and being diagonal in the coordinates and spin.

An ever better one-particle approximation for the grand canonical distribution is the Hartree–Fock one, allowing also for terms non-diagonal in the coordinates and spins. The former simple arguing, however, cannot provide for it, The Hartree–Fock approximation implies a non-diagonal effective potential

$$H^{\mathscr{H}\mathscr{F}} \equiv -\int dx\psi(x)\frac{\hbar^2}{2m}\nabla^2\psi_\sigma(x) + \int dx\int dx'\psi(x)^+U^{\mathscr{H}\mathscr{F}}_{eff}(x,x')\psi(x') + const^{\mathscr{H}\mathscr{F}}$$

defined as

$$U^{\mathscr{H}\mathscr{F}}_{eff}(x,x') \equiv \delta(x-x')\int dy\left(\frac{e^2}{|x-y|}\langle\psi(y)^+\psi(y)\rangle\right) - \frac{e^2}{|x-x'|}\langle\psi(x')^+\psi(x)\rangle$$

with the Hartree Fock constant

$$const^{\mathcal{HF}} \equiv \frac{1}{2}\int dx \int dx' \frac{e^2}{|x-x'|} \times \\ \left[\langle\psi(x)^+\psi(x)\rangle\langle\psi(x')^+\psi(x')\rangle - \langle\psi(x)^+\psi(x')\rangle\langle\psi(x')^+\psi(x)\rangle\right].$$

Again, it may be shown that $\langle H^{\mathcal{HF}}\rangle = \langle H\rangle$ and its time-dependent variant conserves the average energy in the absence of time-dependent external potential. The new Coulomb term containing the non-local average $\langle\psi(x')^+\psi(x)\rangle$ is called Coulomb exchange energy.

In both schemes, one has reduced the many-body problem to a self-consistent one-particle problem. However, for the treatment of some phase transitions with spontaneous symmetry breaking of the particle number conservation (like in the case of Bose condensation or superconductivity we shall discuss in Chap. 7) one needs generalizations of the Hartree–Fock scheme in keeping also particle number non-conserving averages as $\langle\psi(x')\psi(x)\rangle$.

3.2.2 Electron Gas with Coulomb Interactions

As a simple illustration of the Hartree–Fock approximation we consider here a gas of electrons that interact through Coulomb forces, without any periodical potential at zero temperature (ground state). To stabilize the system without losing translational invariance one needs, however, to consider a uniform positive background to replace the positive ions. In the absence of a magnetic field we have spin independence and therefore,

$$\langle\psi_\sigma(\mathbf{r})^+\psi_{\sigma'}(\mathbf{r}')\rangle = \delta_{\sigma,\sigma'}\eta(\mathbf{r}\,\mathbf{r}')$$

while from the homogeneity it follows:

$$\eta(\mathbf{r},\mathbf{r}') \equiv \eta(|\mathbf{r}-\mathbf{r}'|).$$

Then the Hartree–Fock Hamiltonian is

$$H^{\mathcal{HF}} = \sum_{\sigma,\sigma'}\int d\mathbf{r}\int d\mathbf{r}'\psi_\sigma(\mathbf{r})h^{\mathcal{HF}}_{\sigma\sigma'}(\mathbf{r},\mathbf{r}')\psi_{\sigma'}(\mathbf{r}')$$

$$h^{\mathcal{HF}}_{\sigma\sigma'}(\mathbf{r},\mathbf{r}') = \delta_{\sigma\sigma'}\left\{h_0(\mathbf{r},\mathbf{r}') + \delta(\mathbf{r}-\mathbf{r}')\int d\mathbf{r}''\frac{\langle n\rangle e^2}{|\mathbf{r}-\mathbf{r}''|} - \frac{e^2}{|\mathbf{r}-\mathbf{r}'|}\eta(\mathbf{r}-\mathbf{r}')\right\}$$

with

$$h_0(\mathbf{r}, \mathbf{r}') \equiv \delta(\mathbf{r} - \mathbf{r}') \left(-\frac{\hbar^2}{2m} \nabla^2 + V_+ \right).$$

The term

$$V_+ = -\langle n \rangle e^2 \int d\mathbf{r}' \frac{1}{|\mathbf{r}'|}$$

represents the Coulomb potential energy due to the uniform positive background charge and $\langle n \rangle$ is the average density of electrons ($\langle n \rangle = 2\eta(0)$).

Of course, all the homogeneity assumptions are true only in an infinite system and this constant potential energy V_+ is divergent. However, we keep for a while the volume finite and as we shall see, this diverging entity will compensate other inherently diverging entities. Without the positive background contribution, the positive energy of the Coulomb repulsion between the electrons would push them far away, without reaching a stable state.

Due to the homogeneity, the Hartree–Fock eigenfunctions have to be plane waves

$$\phi_{\mathbf{k}}(\mathbf{r}) = \frac{1}{\sqrt{\Omega}} e^{-\imath \mathbf{k}\mathbf{r}}$$

and therefore with the step-like Fermi function $\theta(k_F - |\mathbf{k}|)$ at $T = 0$ the average density is

$$\langle \psi_\sigma(\mathbf{r})^+ \psi_{\sigma'}(\mathbf{r}') \rangle \equiv \delta_{\sigma,\sigma'} \frac{1}{\Omega} \sum_{|\mathbf{k}|<k_F} e^{\imath \mathbf{k}(\mathbf{r}-\mathbf{r}')}.$$

Then the Hartree–Fock eigenvalue equation

$$\left\{ -\frac{\hbar^2}{2m} \nabla^2 - V_+ + \frac{1}{\Omega} \sum_{|\mathbf{k}'|<k_F} \int d\mathbf{r}' \frac{e^2}{|\mathbf{r} - \mathbf{r}'|} \left[2 - e^{\imath(\mathbf{k}-\mathbf{k}')(\mathbf{r}-\mathbf{r}')} \right] \right\} e^{-\imath \mathbf{k}\mathbf{r}} = E_k e^{-\imath \mathbf{k}\mathbf{r}}$$

just defines the energy eigenvalues E_k as

$$E_k = \frac{\hbar^2 k^2}{2m} - \frac{e^2}{\Omega} \sum_{|\mathbf{k}'|<k_F} \int d\mathbf{r}' e^{\imath(\mathbf{k}'-\mathbf{k})(\mathbf{r}-\mathbf{r}')} \frac{1}{|\mathbf{r} - \mathbf{r}'|}.$$

Using the (improper) definition of the Fourier transform of the Coulomb potential by the limit procedure

$$\lim_{\kappa \to 0} \int d\mathbf{r} \; \frac{e^2 e^{-\kappa r}}{r} e^{\imath \mathbf{q}\mathbf{r}} = \frac{4\pi e^2}{q^2}$$

one gets finally

$$E_k = \frac{\hbar^2 k^2}{2m} - \frac{e^2 k_F}{4\pi}\left(2 + \frac{k_F^2 - k^2}{k k_F} \ln\left|\frac{k + k_F}{k - k_F}\right|\right).$$

The total average energy of an electron of the Coulomb electron gas (per spin) differs from that of the free electron gas in the ground state $\frac{3}{5}E_F$

$$\langle E \rangle \equiv \frac{\sum_{|\mathbf{k}|<k_F} E_k}{\sum_{|\mathbf{k}|<k_F} 1} = \frac{3}{5}E_F - \frac{3e^2 k_F}{8\pi}$$

by the "exchange energy" $-\frac{3e^2 k_F}{8\pi}$.
The "exchange correlation function" given by

$$\frac{1}{n}\eta(\mathbf{r}) = \frac{1}{n}\sum_{\sigma} < \psi_\sigma(0)^+ \psi_{\sigma'}(\mathbf{r}) > = \frac{1}{\Omega}\sum_{|\mathbf{k}|<k_F} e^{-\imath \mathbf{k}\mathbf{r}} = -\frac{3}{2(k_F r)^3}(k_F r \cos k_F r - \sin k_F r)$$

defines the probability amplitude to find an electron at the distance $r \equiv |\mathbf{r}|$ as function of the variable $k_F r$ and it looks as in Fig. 3.2.

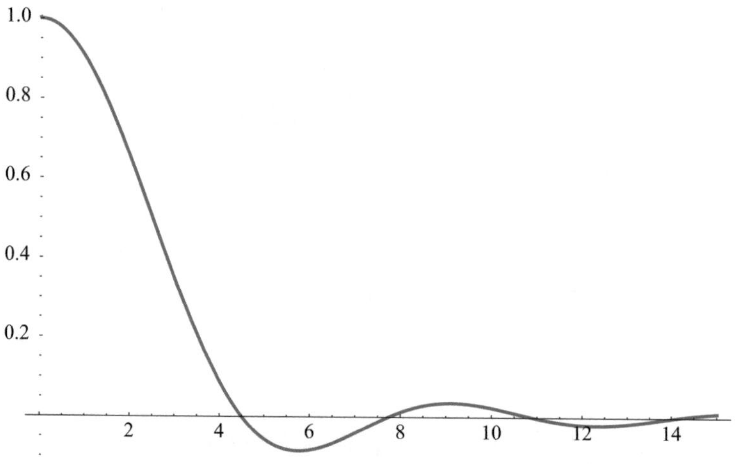

Fig. 3.2 Exchange correlation function

3.2.3 The Electron–Hole Plasma

While the previously described model is used with reference to metals, the model of a Coulomb interacting electron–hole plasma we describe here tries to catch the main many-body aspects in a semiconductor. We have seen that unoccupied electron states in a valence band with negative effective mass act as positively charged particles (holes) with positive effective mass. The negatively charged electrons in the conduction band (from now on just called—electrons, if not otherwise specified) and the positively charged holes interact by Coulomb forces. The Hamiltonian of such a plasma in terms of the second quantized wave functions of electrons and holes

$$\psi_{e,\sigma}(\mathbf{r}) = \sum_{\mathbf{k}} \frac{1}{\sqrt{\Omega}} e^{\imath \mathbf{kr}} a_{e\mathbf{k}\sigma} \;\; ; \quad \psi_{h,\sigma}(\mathbf{r}) = \sum_{\mathbf{k}} \frac{1}{\sqrt{\Omega}} e^{\imath \mathbf{kr}} a_{h\mathbf{k}\sigma}$$

is given by

$$\begin{aligned} H = & \sum_{\mathbf{k},\sigma} (\frac{\hbar^2}{2m_e} k^2 + \frac{1}{2} E_g) a^+_{e,\mathbf{k},\sigma} a_{e,\mathbf{k},\sigma} + \sum_{\mathbf{k},\sigma} (\frac{\hbar^2}{2m_h} k^2 + \frac{1}{2} E_g) a^+_{h,\mathbf{k},\sigma} a_{h,\mathbf{k},\sigma} + E_0 \\ & + \frac{1}{2} \sum_{\sigma,\sigma'} \int d\mathbf{r} \int d\mathbf{r}' \frac{e^2}{|\mathbf{r}-\mathbf{r}'|} \Big\{ \psi_{e,\sigma}(\mathbf{r})^+ \psi_{e,\sigma'}(\mathbf{r}')^+ \psi_{e,\sigma'}(\mathbf{r}') \psi_{e,\sigma}(\mathbf{r}) \\ & + \psi_{h,\sigma}(\mathbf{r})^+ \psi_{h,\sigma'}(\mathbf{r}')^+ \psi_{h,\sigma'}(\mathbf{r}') \psi_{h,\sigma}(\mathbf{r}) \\ & - 2\psi_{e,\sigma}(\mathbf{r})^+ \psi_{e,\sigma}(\mathbf{r}) \psi^+_{h,\sigma'}(\mathbf{r}') \psi_{h,\sigma'}(\mathbf{r}') \Big\} \, . \end{aligned}$$

Here one ignores the Bloch character of the wave functions. The electrons and holes are considered just as free propagating particles. This Hamiltonian conserves the number of electrons N_e as well as of the holes N_h, while the original Coulomb interaction of the conduction and valence band electrons

$$\frac{1}{2} \int d\mathbf{x} \int d\mathbf{x}' \psi(\mathbf{r})^+ \psi(\mathbf{r}')^+ \frac{e^2}{|\mathbf{r}-\mathbf{r}'|} \psi(\mathbf{r}') \psi(\mathbf{r})$$

conserves the total number of electrons in both bands, or otherwise stated it only conserves the difference between the number of electrons and holes $N_e - N_h$. Indeed, the second quantized wave function of conduction and valence band electrons is by definition $\psi_\sigma(\mathbf{x}) = \psi_{e,\sigma}(\mathbf{x}) + \psi_{h,-\sigma}(\mathbf{x})^+$ and therefore, the Coulomb interaction term contains not only conduction band electron and valence band hole charge densities, but also mixed ("exchange") terms.

However, since the Coulomb potential varies slowly within the elementary cell, these terms are small. The matrix element of a function $F(\mathbf{r})$ between Bloch states may be written as a sum of integrals over the elementary cell v

$$\int d\mathbf{r} u_{\mathbf{k}}^{\alpha}(\mathbf{r})^* u_{\mathbf{k}'}^{\beta}(\mathbf{r}) F(\mathbf{r}) = \sum_{\mathbf{R}} \int_{v} d\mathbf{r} u_{\mathbf{k}}^{\alpha}(\mathbf{r})^* u_{\mathbf{k}'}^{\beta}(\mathbf{r}) F(\mathbf{r}).$$

If the function $F(\mathbf{r})$ varies slowly within the elementary cell, then we may replace it by its value in the cell (v), i.e., we get

$$\approx \sum_{\mathbf{R}} F(\mathbf{R}) \int_{v} d\mathbf{r} u_{\mathbf{k}}^{\alpha}(\mathbf{r})^* u_{\mathbf{k}'}^{\beta}(\mathbf{r})$$

and further at small wave vectors

$$\approx \sum_{\mathbf{R}} F(\mathbf{R}) \int_{v} d\mathbf{r} u_{0}^{\alpha}(\mathbf{r})^* u_{0}^{\beta}(\mathbf{r}) = \delta_{\alpha,\beta} \sum_{\mathbf{R}} F(\mathbf{R}) .$$

In the electron–hole plasma model ignored "exchange" terms contain such integrals with $\alpha \neq \beta$ and therefore may be neglected.

Nevertheless, as we shall see later, the interaction with electromagnetic fields (photons) may create or annihilate electron–hole pairs and this is the basic process in the optics of semiconductors.

3.2.4 *Many-Body Perturbation Theory of Solid State*

Beyond the Hartree–Fock approximation one has at disposal just the diagrammatic perturbation theory to improve it. Of course, within this frame one has developed sophisticated methods, but the first question is how to start. In solid state theory, it seems not too clever to consider the motion of the electrons in the field of the rigid ions as the unperturbed problem and to treat the Coulomb interaction as a perturbation. A better way is to start from the Hartree–Fock approximation and treat the difference to the true Hamiltonian as perturbation. According Sect. 3.2.1

$$\begin{aligned} H' = H - H^{\mathscr{HF}} = \frac{1}{2} \int dx \int dx' \frac{e^2}{|x - x'|} \{ & \psi(x)^+ \psi(x')^+ \psi(x') \psi(x) \\ & - 2\langle \psi^+(x') \psi(x') \rangle \psi^+(x) \psi(x) + 2\langle \psi^+(x') \psi(x) \rangle \psi^+(x) \psi(x') \\ & + \langle \psi^+(x) \psi(x) \rangle \langle \psi^+(x') \psi(x') \rangle - \langle \psi^+(x') \psi(x) \rangle \langle \psi^+(x) \psi(x') \rangle \} . \end{aligned}$$

In terms of the Hartree–Fock one-electron eigenstates $\phi_\nu(x)$ the second quantized wave function of the electron in the crystal is

$$\psi(x) = \sum_{\nu} A_\nu \phi_\nu(x)$$

with A_ν being the annihilation operator of the electron state with quantum number ν. This quantum number may be the band index and wave vector of a Bloch state, but we do not need to specify it here. In terms of these operators the Hartree–Fock Hamiltonian is

$$H^{\mathscr{H}\mathscr{F}} = \sum_\sigma \sum_\nu \epsilon_\nu A^+_{\nu,\sigma} A_{\nu,\sigma} + const.^{\mathscr{H}\mathscr{F}}$$

We choose the Hartree–Fock ground state ($T = 0$) for the averages. Also all the states with energy below the Fermi energy $\epsilon_\nu < \epsilon_F$ being occupied, while above it are all empty. The averages themselves are then

$$\langle \psi^+(x)\psi(x')\rangle = \sum_\nu \phi^*_\nu(x)\phi_\nu(x')\theta(\epsilon_F - \epsilon_\nu).$$

Now it is useful to introduce new electron–hole annihilation (and creation) operators $a_{e,\nu}$ (for $\epsilon_\nu > \epsilon_F$) and $a_{h,\nu}$ (for $\epsilon_\nu < \epsilon_F$) by

$$A_\nu = \begin{cases} a_{e,\nu} & (\epsilon_\nu > \epsilon_F) \\ a^+_{h,\tilde{\nu}} & (\epsilon_\nu < \epsilon_F). \end{cases}$$

For the Bloch indices, if $\nu = n, \mathbf{k}, \sigma$ the conjugate quantum number is $\tilde{\nu} = n, -\mathbf{k}, -\sigma$. Within the new description the many-body ground state $|0>$ is the electron–hole vacuum

$$a_{e,\nu}|0>= 0; \quad a_{h,\nu}|0>= 0\,.$$

If one expresses the perturbation H' in terms of the electron–hole creation–annihilation operators and performs all the necessary commutations such that all the electron–hole annihilation operators a_ν are on the right side, while the creation operators a^+_ν are on the left side, then one may see that all the supplementary terms disappear. In other words, the Coulomb terms are equivalent to the "normal ordered" Coulomb interaction

$$H' = \frac{1}{2}\int dx \int dx' \frac{e^2}{|x - x'|} N\left[\psi(x)^+\psi(x')^+\psi(x')\psi(x)\right]. \tag{3.1}$$

Besides the simplicity of the formulation, the normal ordering helps within the many-body adiabatic perturbation theory to simplify the Feynman diagram technique borrowed from the relativistic quantum field theory.

3.2.5 Adiabatic Perturbation Theory

To go beyond the Hartree–Fock self-consistent approximation for a many-body system one has developed perturbation methods borrowed from the relativistic quantum field theory. Without going into all the technical details of these sophisticated techniques we want to describe here their main ideas. One of the reasons for this lies in their importance in solid state calculations, but also in their demystification.

Let us consider, for example, the interaction Hamiltonian 3.1 as a small perturbation. A convenient method to develop a successive expansion in powers of this perturbation is to introduce it slowly adiabatic at an infinite negative time and eliminate it again adiabatic slowly at an infinite positive time, i.e., we replace H' by

$$H'(t) \equiv H' e^{-0|t|},$$

where the notation -0 indicates an infinitesimal parameter that shall be eliminated only after performing all the time integrals. This is necessary, since otherwise the integrals over an infinite time interval are mathematically not well defined.

It is useful then to consider the unitary evolution operator $U(t, t_0)$ in the interaction picture, defined by the Schrödinger equation

$$\imath\hbar \frac{\partial}{\partial t} U(t, t_0) = H'(t)_I U(t, t_0),$$

where the perturbation in the interaction picture $H'(t)_I$ is defined as

$$H'(t)_I \equiv e^{\frac{\imath}{\hbar} H_0 t} H'(t) e^{-\frac{\imath}{\hbar} H_0 t}$$

and of course $U(t_0, t_0) = 1$.

It can be shown that the formal solution of the above equation may be written as a chronological product

$$U(t, t_0) = T \left\{ e^{-\frac{\imath}{\hbar} \int_{t_0}^{t} H'(t')_I dt'} \right\} .$$

The chronological product is understood as an ordering of all the involved terms in the expansion of the exponential according to their increasing time succession.

The central object of the theory is then the S-matrix

$$S \equiv U(\infty, -\infty) .$$

In the case of a time independent perturbation H' one assumes that the ground state of the system evolves by the adiabatic introduction of the perturbation till any finite time to the ground state of the interacting system, apart from a possible phase factor. (This remains a plausible assumption, however, without any proof!) Then the average of any operator $\langle A(t)\rangle$ (at any finite time t) on the ground state of the interacting system is given by

$$\langle A(t)\rangle = \frac{1}{\langle S\rangle_0}\langle U(\infty, t)A(t)_I U(t, -\infty)\rangle_0 = \frac{\langle T\{A(t)_I S\}\rangle_0}{\langle S\rangle_0} \ ,$$

where $A(t)_I$ is the operator A in the interaction picture and the average $\langle\ldots\rangle_0$ is taken over the unperturbed ground state. The factor $\frac{1}{\langle S\rangle_0}$ just takes care of the above mentioned phase factor. In a similar way one may calculate averages of products of two or more time ordered operators on the ground state of the interacting system as

$$\langle T\{A(t_1)B(t_2)\}\rangle = \frac{\langle T\{A(t_1)_I B(t_2)_I S\}\rangle_0}{\langle S\rangle_0} \ .$$

The first important step that enables the calculation of the S-matrix are Wick's theorems that express the terms of any arbitrary order in its expansion in powers of the interaction Hamiltonian as a sum of products of normal ordered creation and annihilation operators with c-number coefficients constructed from products of pairwise contractions (propagators) of the remaining operators. In our case of electrons the propagators are the averages over the unperturbed ground state of the chronological products of the second quantized electron wave functions

$$\overline{\psi(\mathbf{x}, t)\psi(\mathbf{x}', t')^+} = -\imath\langle\{T\{\psi(\mathbf{x}, t)\psi(\mathbf{x}', t')^+\}\rangle_0 \ . \tag{3.2}$$

In the simplest case of a Hamiltonian H_0 describing a homogeneous system this propagator depends actually only on the difference $\mathbf{x} - \mathbf{x}'$ and $t - t'$.

The next step essentially simplifying the calculations are the Feynman diagrams and rules that allow a simple graphical representation of the c-number coefficients and constitute a sort of modular system to assemble complicated structures. These were borrowed from the relativistic quantum field theory. The basic elements are the vertexes corresponding to the interaction Hamiltonian and the lines connecting them that correspond to the particle propagators. With the help of these rules one may perform advanced calculations including summation of certain classes of terms.

The knowledge of the S-matrix allows not only the calculations of different averages of observables, but also more complicated characteristic entities, as the Green functions. These are ground state averages of chronological products of several second quantized wave functions $\psi(\mathbf{x}, t)$ and their conjugates that contain important information about the excitation spectrum above the ground state.

For the sake of illustration let us consider some Feynman diagrams in the theory of Coulomb interacting electrons gas (for simplicity ignoring spin). The Coulomb interaction Hamiltonian in the space of plane waves with cyclic boundary conditions looks in this case as

$$H' = \sum_{\mathbf{p},\mathbf{p}',\mathbf{q}} \frac{2\pi e^2}{q^2} a_{\mathbf{p}}^+ a_{\mathbf{p}'}^+ a_{\mathbf{p}'-\mathbf{q}} a_{\mathbf{p}+\mathbf{q}} \ . \tag{3.3}$$

However, if one starts from the Hartree–Fock approximation (see Sect. 3.2.3) including also the compensating uniform positive background charge

$$H^{\mathcal{HF}} = \sum_{\mathbf{p}} E_{\mathbf{p}} A_{\mathbf{p}}^+ A_{\mathbf{p}} \ ,$$

then the second quantized electron wave functions are defined in terms of the HF creation–annihilation operators as the second quantized wave function is

$$\psi(\mathbf{x}) = \sum_{\mathbf{p}} \frac{1}{\sqrt{\Omega}} e^{\imath \mathbf{p}\mathbf{x}} A_{\mathbf{p}}$$

and the perturbation is the normal ordered Hamiltonian

$$\mathcal{H}' = \sum_{\mathbf{p},\mathbf{p}',\mathbf{q}} \frac{2\pi e^2}{q^2} N\{A_{\mathbf{p}}^+ A_{\mathbf{p}'}^+ A_{\mathbf{p}'-\mathbf{q}} A_{\mathbf{p}+\mathbf{q}}\} \tag{3.4}$$

with respect to the new creation and annihilation operators of "electrons above" and "holes below" the ground state:

$$A_{\mathbf{p}} = \begin{cases} \alpha_{e,\mathbf{p}} & (E_{\mathbf{p}} > E_F) \\ \alpha_{h,-\mathbf{p}}^+ & (E_{\mathbf{p}} < E_F). \end{cases}$$

The unperturbed ground state is correspondingly the "electron"–"hole" vacuum $|0\rangle$

$$\alpha_{e,\mathbf{p}}|0\rangle = 0 \qquad \alpha_{h,\mathbf{p}}|0\rangle = 0 \ .$$

The unperturbed propagator in the 4-dimensional Fourier space is given by the simple expression (HF Green function)

$$G_{HF}(\mathbf{p},\omega) = \frac{1}{\hbar\omega - E_{\mathbf{p}} + \imath 0 sign(E(\mathbf{p}) - E_F)} \tag{3.5}$$

and will be represented in the Feynman diagrams with a solid line. In the case of the crystal with Bloch bands the expression of the propagator of course is more complicated.

Equation 3.4 defines the vertex in the language of Feynman diagrams shown in Fig. 3.3 with two in-going electron lines and two out-going ones, exchanging a Coulomb dashed line carrying the momentum $\mathbf{q}$ according to the interaction Hamiltonian Eq. 3.4.

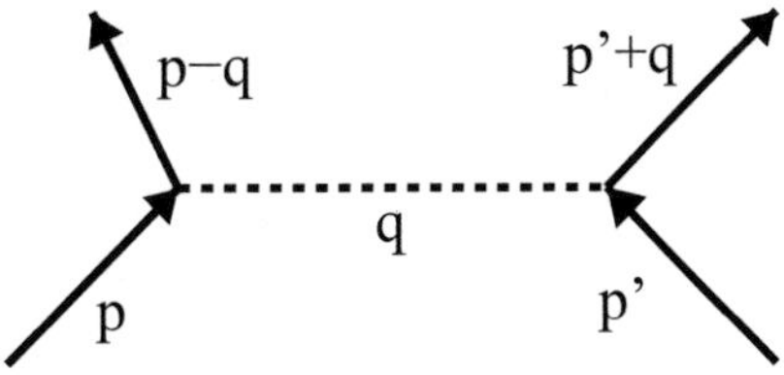

Fig. 3.3 The Coulomb vertex

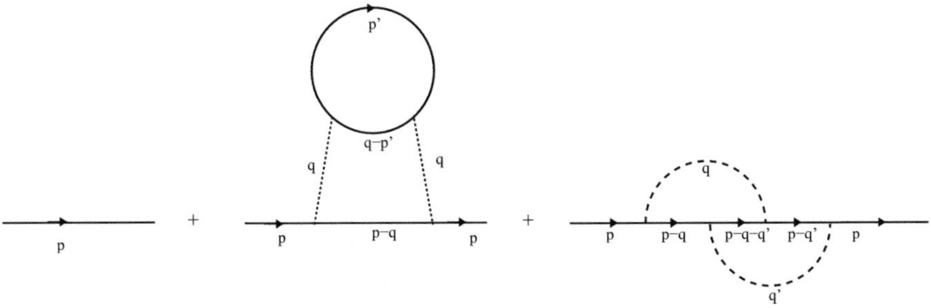

Fig. 3.4 Zeroth and second order $\Sigma_2(\mathbf{p}, \omega)$ "self-energy" terms for the Green function starting from HF

An important aspect of Wick's theorems is that no contractions should be taken inside normal ordered interaction Hamiltonians, like that of Eqs. 3.1 and 3.4. Therefore the perturbation expansion in our case starts already with the second order terms shown in Fig. 3.4.

The funny thing about the Feynman diagrams is that any connected part, once calculated may be inserted anywhere as many times as one desires leading to a correction of respective order. This is the modular principle we hinted before. For example, if one considers all corrections with multiple insertions of the second order "self-energy" diagram $\Sigma_2(\mathbf{p})$, they sum up to the geometrical series and give rise to the approximate Green function

$$G(\mathbf{p}, \omega) = \frac{1}{\hbar\omega - E(\mathbf{p}) - \Sigma_2(\mathbf{p}, \omega) + \imath 0 sign(E(\mathbf{p}) - E_F)} \ .$$

If, however, one starts from the free Hamiltonian (not from the HF-one), then the unperturbed propagator is

$$G_0(\mathbf{p}, \omega) = \frac{1}{\hbar\omega - \frac{\hbar^2 k^2}{2m} + \imath 0 sign(\frac{\hbar^2 k^2}{2m} - E_F)} \qquad (3.6)$$

and also first order diagrams contribute to the "self-energy." These are shown in Fig. 3.5. The first correction is divergent, but actually may be omitted, since it is compensated by the Coulomb interaction with the homogeneous positive charge. The remaining one is automatically included in the HF-approximation.

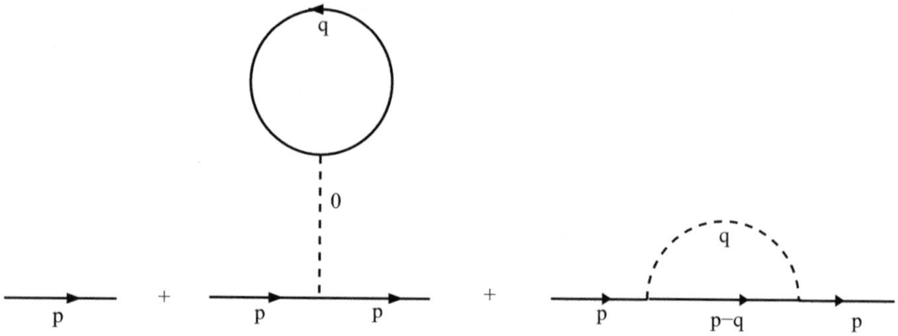

Fig. 3.5 First order "self-energy" terms of the Green function starting from free electrons

Besides the above described diagram technique for averages over the ground state, there are extensions of this formalism for averages over the grand canonical distribution (Matsubara diagrams) or for averages on non-equilibrium states (Keldysh diagrams) we do not touch here at all.

Now, a few words about the connection of the many-body S-matrix theory to that of the relativistic quantum field theory. The primary interest in the later is the computation of the transition probability amplitudes given by the matrix elements of the S matrix between states having a finite number of free particles and the ground state is the true vacuum.

One has to stress that the similarity of the two theories is rather limited and from the strict mathematical point of view both stay on a shaky ground. Besides the open questions about convergence of the series, both are plagued by diverging integrals at high momenta ("ultraviolet divergences"). The ultraviolet problems are tackled, however, in quite a different way. In the non-relativistic many-body theories one restricts the integration domains by physically relevant momenta appropriate for the system one is modeling. In the case of the electrons in a crystal this is the Brillouin zone. Such kind of theories are called "cut-off" theories. Here the "bare" (unrenormalized, i.e., in the absence of the interaction) electron and hole masses $m_e,\ m_h$, as well as the electron charge e are the experimental ones. On the contrary, in the quantum field theories the divergences are eliminated by the so-called renormalization procedure admitting that actually the "bare" (input) parameters are infinite and the renormalized ones are fixed to give the finite experimental ones.

Chapter 4
Phonons

The ions forming the attractively acting lattice are not rigid. At least their oscillations around the rigid lattice positions have to be taken into account. These oscillations may be quantized and interpreted as bosonic particles called acoustical or optical phonons with typical spectra. Even a classical description of the phonons gives rise to important predictions as the Lyddane–Sachs–Teller formula. In a quantum mechanical approach to optical transitions assisted by phonons one may understand the Franck–Condon effect.

4.1 Lattice Oscillations

Until now we considered the ions as fixed positive point-like charges, positioned at the sites $\mathbf{R} + \boldsymbol{\xi}_s$ $(s = 1 \ldots S)$, where $\mathbf{R}$ is a vector of the Bravais lattice, while $\boldsymbol{\xi}_s$ shows the position of a given ion within the elementary cell containing S ions. Actually, the ions are free to move away from their equilibrium positions with deviations $\mathbf{u}(\mathbf{R}, s)$ (here still classical!). For small deviations one may develop the potential energy $\mathscr{U}$ of the lattice in a power series of these deviations. Up to a constant defining the ground state energy, the lowest term in this series must be quadratic, since the equilibrium corresponds to a stable minimum. Thus

$$\mathscr{U} = \frac{1}{2} \sum_{\mathbf{R},s} \sum_{\mathbf{R}',s'} \Phi^{\mu\nu}_{ss'}(\mathbf{R} - \mathbf{R}') u^{\mu}(\mathbf{R}, s) u^{\nu}(\mathbf{R}' s') + \ldots ,$$

where $\Phi^{\mu\nu}_{ss'}(\mathbf{R} - \mathbf{R}')$ is a real symmetric

$$\Phi^{\mu\nu}_{ss'}(\mathbf{R}) = \Phi^{\nu\mu}_{s's}(-\mathbf{R})$$

L. A. Bányai, *A Compendium of Solid State Theory*,
https://doi.org/10.1007/978-3-030-37359-7_4

positive defined matrix ($\Phi \geq 0$ means non-negative eigenvalues).
Since a constant shift does not change the energy

$$\sum_{\mathbf{R},s} \Phi^{\mu\nu}_{ss'}(\mathbf{R}) = \sum_{\mathbf{R},s} \Phi^{\mu\nu}_{s's}(\mathbf{R}) = 0.$$

One gets the classical Lagrange function $L = \mathscr{T} - U$ of the lattice by considering also the kinetic energy of the lattice

$$\mathscr{T} = \frac{1}{2} \sum_{\mathbf{R},s} m_s \dot{\mathbf{u}}(\mathbf{R}, s)^2,$$

where m_s is the mass of the ion designed with the index s.
In terms of discrete Fourier transforms

$$\mathbf{u}(\mathbf{R}, s) = \frac{v}{(2\pi)^3} \int_{BZ} d\mathbf{q} e^{\imath \mathbf{q}\mathbf{R}} \tilde{\mathbf{u}}_s(\mathbf{q}); \qquad \tilde{\Phi}^{\mu\nu}_{ss'}(\mathbf{q}) = \frac{v}{(2\pi)^3} \int_{BZ} d\mathbf{q} e^{\imath \mathbf{q}\mathbf{R}} \Phi^{\mu\nu}_{ss'}(\mathbf{R})$$

retaining only the quadratic terms of the potential energy we have

$$L = \frac{v}{2(2\pi)^3} \int_{BZ} d\mathbf{q} \left\{ \sum_{s=1}^{S} \sum_{\mu} m_s \dot{\tilde{u}}^{\mu}_s(\mathbf{q})^* \dot{\tilde{u}}^{\mu}_s(\mathbf{q}) - \sum_{s,s'=1}^{S} \sum_{\mu,\nu} \tilde{u}^{\mu}_s(\mathbf{q})^* \tilde{\Phi}(\mathbf{q})^{\mu\nu}_{ss'} \tilde{u}^{\nu}_{s'}(\mathbf{q}) \right\}$$

with

$$\tilde{\Phi}(\mathbf{q})^{\mu\nu *}_{ss'} = \tilde{\Phi}(\mathbf{q})^{\nu\mu}_{s's}; \qquad \tilde{\Phi}(\mathbf{q})^{\mu\nu *}_{ss'} = \tilde{\Phi}(-\mathbf{q})^{\mu\nu}_{ss'}; \qquad \tilde{\Phi}(\mathbf{q}) \geq 0.$$

Absorbing the mass factor by

$$\tilde{\eta}^{\mu}_s(\mathbf{q}) \equiv \sqrt{m_s} \tilde{u}^{\mu}_s(\mathbf{q}); \qquad \tilde{M}(\mathbf{q})^{\mu\nu}_{ss'} \equiv \frac{1}{\sqrt{m_s m_{s'}}} \tilde{\Phi}(\mathbf{q})^{\mu\nu}_{ss'}$$

we get

$$L = \frac{v}{2(2\pi)^3} \int_{BZ} d\mathbf{q} \left\{ \sum_{s=1}^{S} \sum_{\mu} \dot{\tilde{\eta}}^{\mu}_s(\mathbf{q})^* \dot{\tilde{\eta}}^{\mu}_s(\mathbf{q}) - \sum_{s,s'=1}^{S} \sum_{\mu,\nu} \tilde{\eta}^{\mu}_s(\mathbf{q})^* \tilde{M}(\mathbf{q})^{\mu\nu}_{ss'} \tilde{\eta}^{\nu}_{s'}(\mathbf{q}) \right\},$$

where the new matrix $\tilde{M}$ has the same properties of symmetry, realness, and positiveness as $\tilde{\Phi}$.
This whole quadratic form may be brought to a diagonal one with the transformation that diagonalizes the potential energy, with positive eigenvalues $\omega_\lambda(\mathbf{q})^2$ and one may see that the Lagrangian describes a sum of oscillators of unit mass with eigenfrequencies $\omega_\lambda(\mathbf{q}) = \omega_\lambda(-\mathbf{q})$.

The quantization of the lattice corresponds therefore to the quantization of these oscillators. As it is well known, this leads to a quantum mechanical Hamiltonian that may be formulated in terms of the creation and annihilation operators $b^{\dagger}_{\lambda\mathbf{q}}$, $b_{\lambda\mathbf{q}}$ and the respective oscillator frequencies $\omega_{\lambda}(\mathbf{q})$. It is convenient to impose cyclical boundary conditions in order to deal with a discrete spectrum having the proper number of degrees of freedom. Then the quantized lattice Hamiltonian is

$$H = \sum_{\mathbf{q}\in BZ} \sum_{\lambda} \hbar\omega_{\lambda\mathbf{q}} b^{\dagger}_{\lambda\mathbf{q}} b_{\lambda\mathbf{q}}$$

with the bosonic commutation relations

$$\left[b_{\lambda\mathbf{q}}, b_{\lambda'\mathbf{q}'}\right] = 0$$

$$\left[b_{\lambda\mathbf{q}}, b^{\dagger}_{\lambda'\mathbf{q}'}\right] = \delta_{\lambda,\lambda'}\delta_{\mathbf{q},\mathbf{q}'}.$$

The quantized deviations from the equilibrium positions at their turn are

$$\mathbf{u}(\mathbf{R}, s) = \sum_{\mathbf{q}} \sum_{\lambda} e^{\imath\mathbf{q}\mathbf{R}} \sqrt{\frac{\hbar}{2m_s\omega_{\lambda,\mathbf{q}}}} \boldsymbol{\chi}_s^{(\lambda)}(\mathbf{q}) \left(b_{\lambda,\mathbf{q}} + b^{\dagger}_{\lambda,-\mathbf{q}}\right),$$

where $\boldsymbol{\chi}_s^{(\lambda)}(\mathbf{q})$ are the orthonormalized eigenfunctions of the Matrix $\tilde{M}$

$$\sum_{\nu,s'} \tilde{M}(\mathbf{q})_{s,s'} \chi_{s'}^{\nu(\lambda)}(\mathbf{q}) = \omega^2_{\lambda,\mathbf{q}} \chi_s^{\mu(\lambda)}(\mathbf{q}); \ (\lambda = 1, \ldots 3S).$$

Due to the invariance against a common translation discussed before it follows that

$$\sum_{s'=1}^{S} \tilde{M}(0)^{\mu\nu}_{s,s'} \sqrt{m_{s'}} = 0$$

and

$$\sum_{s=1}^{S} \sqrt{m_s} \tilde{M}(0)^{\mu\nu}_{ss'} = 0.$$

While the first equation shows that there are at least 3 eigenfrequencies that vanish at $\mathbf{q} = 0$, called acoustical modes, the second equation shows that if an eigenvalue does not vanish in the origin, then the eigenstate must fulfill the relation

$$\sum_{s=1}^{S} \sqrt{m_s} \boldsymbol{\chi}_s^{(\lambda)}(0) = 0; \qquad (\lambda = 4, \ldots 3S - 3)$$

or

$$\sum_{s=1}^{S} m_s \mathbf{u}_s^{(\lambda)}(0) = 0; \qquad (\lambda = 4, \ldots 3S - 3).$$

This means that the center of mass by these oscillations, called optical modes, remains unchanged.

4.2 Classical Continuum Phonon-Model

In applications, it is useful to use a simplified continuum model for phonons with a simple phonon spectrum. The local deviations $\mathbf{u}(\mathbf{r})$ are defined then in every space point $\mathbf{r}$.
The prototype classical Lagrange function for acoustical phonons is

$$L_{acc} = \frac{m}{2v} \int d\mathbf{r} \sum_{\mu=1}^{3} \left\{ \dot{u}_\mu(\mathbf{r})^2 + c^2 \sum_{\nu=1}^{3} (\partial_\nu u_\mu(\mathbf{r}))^2 \right\}$$

or in Fourier transforms

$$L_{acc} = \frac{m}{2} \int d\mathbf{q} \sum_{\mu=1}^{3} \left\{ \dot{\tilde{u}}_\mu(\mathbf{q})^* \dot{\tilde{u}}_\mu(\mathbf{q}) + c^2 q^2 \tilde{u}_\mu(\mathbf{q})^* \tilde{u}_\mu(\mathbf{q}) \right\}.$$

Obviously, here a linear acoustical phonon spectrum $\omega_{ac}(\mathbf{q}) = cq$ was assumed (with c being here the sound velocity in the medium).
To model optical phonons, one considers the classical Lagrange function

$$L_{opt} = \frac{m}{2v} \int d\mathbf{r} \sum_{\mu=1}^{3} \left\{ \dot{\tilde{u}}_\mu(\mathbf{r})^2 - \omega_0^2 \tilde{u}_\mu(\mathbf{r}))^2 \right\}$$

or in Fourier transforms

$$L_{opt} = \frac{m}{2} \int d\mathbf{q} \sum_{\mu=1}^{3} \left\{ \dot{\tilde{u}}_\mu(\mathbf{q})^* \dot{\tilde{u}}_\mu(\mathbf{q}) - \omega_0^2 \tilde{u}_\mu(\mathbf{q})^* \tilde{u}_\mu(\mathbf{q}) \right\}.$$

Here the optical phonon spectrum was taken just to be constant $\omega_{opt}(\mathbf{q}) = \omega_0$. In both cases one must take into account that the total number of degrees of freedom (per volume!) should correspond to that of the crystal and therefore one cuts off the wave vectors by the Debye wave vector ($q < q_{Debye}$) defined by

$$\frac{4\pi}{3} q_{Debye}^{3} \equiv \frac{1}{v}$$

with v being the volume of the elementary cell.

4.2.1 Optical Phonons in Polar Semiconductors

The optical lattice deviations lead to local electric dipoles. This local polarization in the classical continuum model is given by

$$\mathbf{P}(\mathbf{r}) = \frac{\kappa e}{v} \mathbf{u}(\mathbf{r}),$$

where κ is the polarizability of the elementary cell. The corresponding polarization charge density is

$$\rho_{pol}(\mathbf{r}) = -\nabla \cdot \mathbf{P}(\mathbf{r}) = -\frac{\kappa e}{v} \nabla \cdot \mathbf{u}(\mathbf{r})$$

and the Poisson equation in the presence of a stationary external charge density $\rho_{ext}(\mathbf{r})$ looks as

$$\epsilon_\infty \nabla^2 V(\mathbf{r}) = -4\pi \left(\rho_{pol}(\mathbf{r}) + \rho_{ext}(\mathbf{r})\right)$$

with ϵ_∞ being the dielectric constant due to the electronic background. Alternatively, for low frequencies it will look as

$$\epsilon_0 \nabla^2 V(\mathbf{r}) = -4\pi \rho_{ext}(\mathbf{r}),$$

where ϵ_0 is the total dielectric constant. The notation stems from the assumption that at high frequencies only the light electrons contribute to the dielectric properties, while for low frequencies both the electrons and ions contribute. As we shall see, relating the polarization charge density to the potential allows for the identification of the total dielectric constant.

In the inhomogeneous situation due to the presence of an external charge density the Lagrange function contains supplementary Coulomb terms

$$L = \frac{m}{2v} \int d\mathbf{r} \sum_{\mu=1}^{3} \left\{ \dot{u}^{\mu}(\mathbf{r})^2 - \omega_0^2 u^{\mu}(\mathbf{r}))^2 \right\}$$
$$- \frac{1}{\epsilon_\infty} \int d\mathbf{r} \int d\mathbf{r}' \frac{1}{|\mathbf{r} - \mathbf{r}'|} \left(\frac{1}{2} \nabla \mathbf{P}(\mathbf{r}) \nabla' \mathbf{P}(\mathbf{r}') - \nabla \mathbf{P}(\mathbf{r}) \rho_{ext}(\mathbf{r}') \right).$$

Now let us split the local deviations $\mathbf{u}(\mathbf{r})$ into longitudinal and transverse modes

$$\mathbf{u}(\mathbf{r}) = \mathbf{u}_l(\mathbf{r}) + \mathbf{u}_t(\mathbf{r})$$

$$\mathbf{u}_l(\mathbf{r}) = -\nabla \int d\mathbf{r}' \frac{\nabla' \mathbf{u}(\mathbf{r}')}{4\pi |\mathbf{r} - \mathbf{r}'|}; \qquad \nabla \mathbf{u}_t = 0; \quad \nabla \mathbf{u}_l = \nabla \mathbf{u}.$$

One gets the equations of motion for these modes through the Euler–Lagrange equations by using the identity

$$\nabla^2 \frac{1}{|\mathbf{r}|} = -4\pi\delta(\mathbf{r})$$

and they look as

$$\frac{\partial^2}{\partial t^2} \mathbf{u}_t = -\omega_{TO}^2 \mathbf{u}_t$$

$$\frac{\partial^2}{\partial t^2} \mathbf{u}_l = -\omega_{LO}^2 \mathbf{u}_l + \frac{\kappa e}{m} \mathbf{E}_{ext}$$

with

$$\omega_{TO}^2 \equiv \omega_0^2; \qquad \omega_{LO}^2 = \omega_0^2 + \frac{4\pi\kappa^2 e^2}{\epsilon_\infty v m}.$$

In a stationary regime (thermal equilibrium) in the presence of a stationary external charge density it follows:

$$\mathbf{u}_l = \frac{\kappa e}{m\omega_{LO}^2} \mathbf{E}_{ext}$$

and

$$\rho_{pol} = -\frac{\kappa^2 e^2}{v m \omega_{LO}^2} \nabla \mathbf{E}_{ext} = -\frac{4\pi\kappa^2 e^2}{v m \omega_{LO}^2 \epsilon_\infty} \rho_{ext}; \qquad (\epsilon_\infty \nabla \mathbf{E}_{ext} = 4\pi\rho_{ext}).$$

If one inserts this result into the Poisson equation, one may identify

$$\frac{\epsilon_\infty}{\epsilon_0} = 1 - \frac{4\pi\kappa^2 e^2}{v m \omega_{LO}^2 \epsilon_\infty} = \frac{1}{\omega_{LO}^2} \left(\omega_{LO}^2 - \frac{4\pi\kappa^2 e^2}{\epsilon_\infty v m} \right)$$

and one gets the Lyddane–Sachs–Teller relationship

$$\frac{\omega_{LO}}{\omega_{TO}} = \sqrt{\frac{\epsilon_0}{\epsilon_\infty}}.$$

Now let us introduce an arbitrary external time-dependent electromagnetic field $\mathbf{E_{ext}}(t)$. Then the equation of motion looks as

$$\frac{\partial^2}{\partial t^2}\mathbf{u}(t) = -\omega_0^2\mathbf{u}(t) + \frac{\kappa e}{m}\mathbf{E}(t) \quad ,$$

whereby the electric field $\mathbf{E}(t)$ includes also the longitudinal field produced by the dipoles

$$\mathbf{E}(\mathbf{x}, t) = \mathbf{E}_{ext}(\mathbf{x}, t) + \nabla \int d\mathbf{x}' \frac{\nabla' P(\mathbf{x}', t)}{|\mathbf{x} - \mathbf{x}'|}.$$

Above we took into account that the interaction energy of our dipoles with the total electric field may be brought to the form $-\frac{\kappa e}{v}\int d\mathbf{x}\mathbf{u}(\mathbf{x})\mathbf{E}(\mathbf{x}, t)$. After a Fourier transformation in the time variable t it follows for the frequency dependent susceptibility for both modes

$$\chi_{T,L}(k, \omega) = \frac{\alpha^2 e^2}{vm}\frac{1}{\omega_0^2 - \omega^2},$$

or for the frequency dependent dielectric function

$$\epsilon(\omega)_{L,T} = \epsilon_\infty\left(1 + \frac{4\pi\alpha^2 e^2}{\epsilon_\infty vm}\frac{1}{\omega_0^2 - \omega^2}\right) = \epsilon_\infty\frac{\omega_{LO}^2 - \omega^2}{\omega_{TO}^2 - \omega^2}.$$

4.2.2 Optical Eigenmodes

Starting from the above dielectric function we look now for possible electromagnetic eigenoscillations in such a medium. Since the system is homogeneous and isotropic, by using Fourier transforms it is useful to split the electromagnetic fields in their longitudinal, respectively, transverse parts. The Maxwell equations for the magnetic $\mathbf{B}$, respectively, electric field $\mathbf{E}$ and the polarization $\mathbf{P}$ look in their components as

$$\begin{aligned}
B(k, \omega)_L &= 0 \\
B(k, \omega)_T &= \frac{ck}{\omega}E(k, \omega)_T \\
\imath k\epsilon_\infty E(k, \omega)_L &= -4\pi\imath k P(k, \omega)_L + 4\pi\rho^{ext}(k, \omega)
\end{aligned}$$

$$(k^2 - \epsilon_\infty \frac{\omega^2}{c^2}) E(k, \omega)_T = \frac{4\pi\omega^2}{c^2} P(k, \omega)_T - \frac{\imath 4\pi\omega}{c^2} j^{ext}(k, \omega)_T.$$

With

$$\mathbf{P}(\mathbf{k}, \omega) = \frac{1}{4\pi} \left(\epsilon(\mathbf{k}, \omega) - \epsilon_\infty\right) \mathbf{E}(\mathbf{k}, \omega)$$

and the previously deduced dielectric function it follows that in the absence of external sources the homogeneous Maxwell equations still have non-vanishing solutions. For $\omega = \omega_{LO}$ there is a non-trivial longitudinal solution for any $k = |\mathbf{k}|$, while non-trivial transverse solutions exist only for ω, k pairs that are solutions of the equation

$$k^2 - \epsilon_\infty \frac{\omega^2}{c^2} \frac{\omega_{LO}^2 - \omega^2}{\omega_{TO}^2 - \omega^2} = 0$$

shown in Fig. 4.1. Such transverse fields are propagating mixed photon and LO-phonon modes.

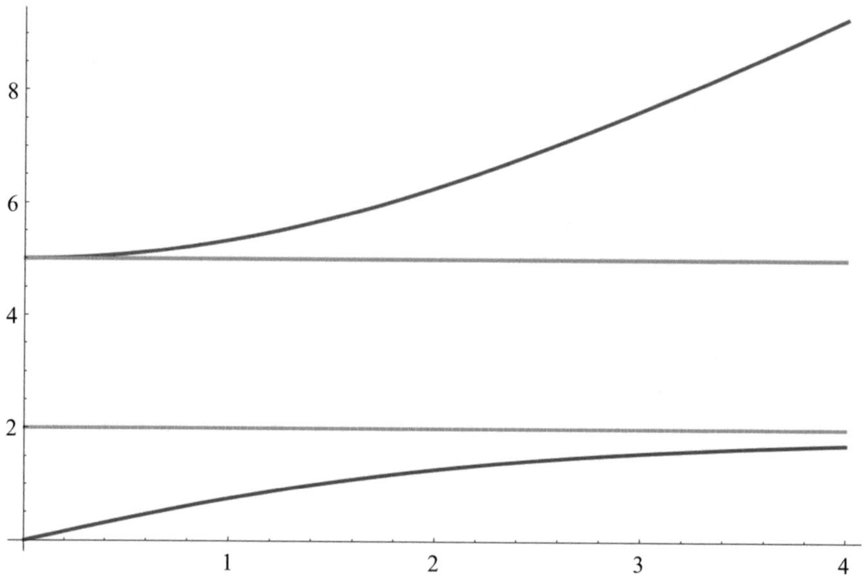

Fig. 4.1 Optical eigenmodes

4.2.3 The Electron–Phonon Interaction

4.2.3.1 The Franck–Condon Effect

Let us suppose that one has two localized electronic states of energies ε_1 and ε_2 on the same site and one of them is electrically neutral while the other one is electrically charged. In a polar semiconductor, this local charge density is coupled to the optical phonons. The energy of the system of electron and phonons (seen as a classical oscillator with coordinate Q) may be characterized schematically by the potential energies of the two states

$$E_1(Q) = \varepsilon_1 + \frac{1}{2}m(\dot{Q}^2 + \omega_{LO}^2 Q^2)$$

$$E_2(Q) = \varepsilon_2 + \frac{1}{2}m(\dot{Q}^2 + \omega_{LO}^2 Q^2) - gQ.$$

The energy of the second state may be rewritten as

$$E_2(Q) = \varepsilon_2 + \frac{1}{2}m\left(\dot{Q}^2 + \omega_{LO}^2(Q - Q_0)^2 + \omega_{LO}^2 Q_0^2\right),$$

where $m\omega_{LO}^2 Q_0 = g$. This means that the potential energy of the phonons gets shifted.

Now by photon absorption the system may undergo a transition from the minimum of the lower lying state on the higher branch outside the minimum of that branch. (The photon momentum may be neglected by these optical transitions!) Thereafter, follows a thermal relaxation (by emission of acoustical phonons) into the minimum of the upper branch and only later may follow a slower photon emission onto the lower branch again outside the minimum. This is the essence of the Franck–Condon effect showing a difference between the absorption and emission spectra (Stokes shift). It is illustrated in Fig. 4.2.

4.2.3.2 The Quantized Interaction of Electrons with Phonons

The interaction Hamiltonian of the electrons with the phonons is considered to be linear in the quantized lattice deviations, conserving the number of electrons and in the continuum model conserving also the momenta

$$H_{int} = \frac{1}{\sqrt{\Omega}} \sum_{\mathbf{q}} g_{\mathbf{q}} a_{\mathbf{k}}^{+} a_{\mathbf{k}-\mathbf{q}} (b_{\mathbf{q}} + b_{-\mathbf{q}}^{+}).$$

Here Ω is the volume and the discretized wave vector $\mathbf{q}$ is assumed to be smaller than the Debye wave vector $q < q_D$.

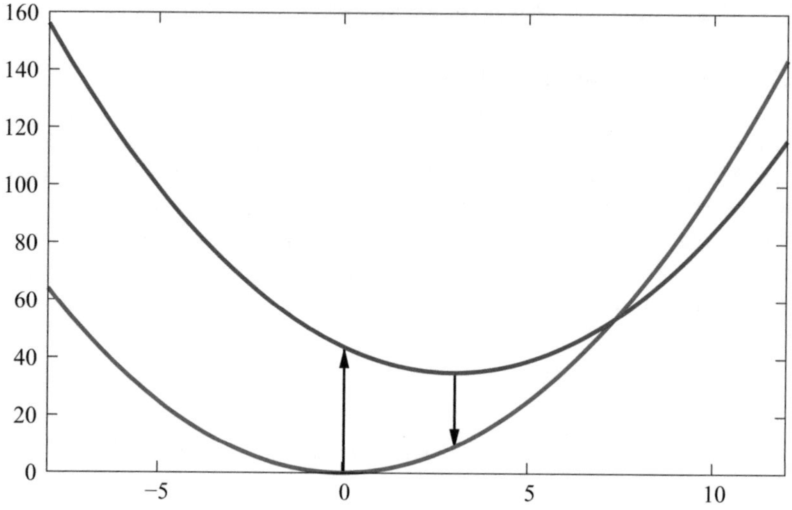

Fig. 4.2 Stokes shift

In the case of the optical phonons, analogously to the discussion in the frame of the classical continuum model one starts from the Coulomb interaction between the electron (or hole) charge density

$$\rho(\mathbf{x}) \equiv \pm e\psi^{+}(\mathbf{x})\psi(\mathbf{x})$$

and the polarization charge density due to the optical deformation

$$-\frac{1}{\epsilon_\infty}\int d\mathbf{x}\int d\mathbf{x}'\frac{\nabla\mathbf{P}(\mathbf{x})\rho(\mathbf{x}')}{|\mathbf{x}-\mathbf{x}'|}.$$

Inserting the expression of the polarization in terms of the quantized lattice deviations one gets

$$g_q^2 = \alpha\frac{4\pi\hbar\,(\hbar\omega_0)^{3/2}}{(2m_e)^{1/2}\,q^2},$$

where instead of the coefficient $\frac{4\pi e^2\kappa^2}{vm\omega_{LO}^2\epsilon_\infty}$ one has introduced the dimensionless constant

$$\alpha = \frac{e^2}{\hbar}\left(\frac{m_e}{2\hbar\omega_0}\right)^{1/2}\left(\frac{1}{\epsilon_\infty}-\frac{1}{\epsilon_0}\right).$$

The presence of the electron mass here is spurious. The coupling constant g_q itself does not depend on the electron mass.

In the case of the acoustic phonons one considers mostly the deformation potential model. This starts from the assumption that a slowly varying deformation in an isotropic medium causes a local variation of the band gap

$$\delta E_g(\mathbf{x}) \sim \nabla \mathbf{u}(\mathbf{x}).$$

This leads to

$$g_{\mathbf{q}} = G\sqrt{\hbar \omega_{\mathbf{q}}},$$

where $\omega_{\mathbf{q}} = c|\mathbf{q}|$ and G is a constant specific for the considered crystal. Of course, both models are highly idealized, but nevertheless, quite successful in predicting experimental phenomena.

Chapter 5
Transport Theory

The understanding of transport phenomena transcends the quantum mechanical treatment of isolated systems. Irreversibility and dissipation due to the interaction with another macroscopic system (bath) play here a central role. For their understanding a simple classical, but exactly solvable model of an electron in the presence of a d.c. electric field interacting with classical phonons is helpful. On the other hand, for all practical purposes an intuitive, however, mathematically less sound treatment serves by introducing irreversibility "by hand" in the Boltzmann, Master or rate equations. In this frame the conductivity of electrons scattered on a thermal bath is discussed. Two basic kinds of conduction are considered. That of scattering of freely moving band electrons, typical for metals and semiconductors, as well as that of localized electrons hopping from site to site, typical for disordered systems, like impurity states or amorphous semiconductors. The transverse magnetoresistance in ultra-high magnetic fields also may be understood within this last model. We discuss shortly also the general theory of kinetic coefficients.

5.1 Non-Equilibrium Phenomena

Most experiments on solid materials concern their electrical, optical, thermal, and mechanical properties. Under such an external influence the solid system is brought into a non-equilibrium state. This may be a stationary state or a transitory time-dependent evolution towards equilibrium. Classical or quantum mechanics of finite closed systems being time reversal invariant cannot properly describe such irreversible processes. The origin of the irreversibility of real physical systems lies in the fact that they are actually open systems connected with some thermostat to

L. A. Bányai, *A Compendium of Solid State Theory*,
https://doi.org/10.1007/978-3-030-37359-7_5

which they transfer (dissipate) energy. In other words, they are the parts of a bigger system. The reversibility even of isolated systems may be, however, pushed far away in time due to the macroscopic size of the system. Therefore, the thermodynamic limit plays an essential role in the treatment of non-equilibrium phenomena.

The traditional theoretical treatment of non-equilibrium phenomena starts, however, directly by assuming some simple, intuitive assumptions leaving their proof apart. In the classical non-quantum mechanical treatment such a useful tool is the Boltzmann equation for the statistical evolution of classical particles having a given velocity and position. We have already discussed that electrons in a periodical lattice may be also assimilated with free electrons having an effective mass. In this sense, one may use this model. On the other hand, one may encounter situations, where the quantum mechanical properties are essential. In that case Master and rate equations are used to describe the evolution due to transitions between quantum mechanical states.

5.2 Classical Solvable Model of an Electron in a d.c. Electric Field Interacting with Phonons

For a better understanding of the origin of irreversibility we describe here a classical solvable model of an electron interacting with classical continuum LO/LA phonons in the presence of a d.c. electric field.

The phenomenological Drude model with a damping term in the electron equation

$$\ddot{x} = \mathcal{E} - \dot{x}/\tau$$

violates obviously time reversal invariance, but describes both stationary motion of the electron in the presence of the field as well as its damping in its absence.

The question is whether the interaction of the electron with a phonon reservoir could lead also to such a behavior? The possibility exists, since although the whole system (electron plus phonons) is time reversal invariant, this is not valid for the electron alone. The electron may lose energy to the phonons and it may happen that this energy does not return in any finite time from the infinite number of phonon degrees of freedom.

The quantum mechanical description of such a complicated system cannot be solved exactly; however, the classical version is exactly solvable and this may well illustrate the origin of the dissipation and irreversibility of this open system (here one electron interacting with the continuum degrees of freedom of phonons).

The Lagrange function of the two kinds of phonon displacements fields $\mathbf{u}(\mathbf{x})$ defined on the continuum were already described in Chap. 4. The Lagrange function of the electron of mass m and charge e in a dc electric field $\mathbf{E}$ is

$$L_{el} = \frac{m\dot{\mathbf{r}}^2}{2} + e\mathbf{E}\mathbf{r}.$$

According to Sect. 4.2.3.2 both phonons couple to the electron by their longitudinal part $d(\mathbf{x}) = \nabla \mathbf{u}(\mathbf{x})$ in the interaction Lagrange function

$$L_{int} = \int d\mathbf{x} V(\mathbf{x} - \mathbf{r}) d((\mathbf{x}).$$

The potential $V(\mathbf{x})$ we define by its Fourier Transform,

$$V(\mathbf{x}) = \int d\mathbf{q} e^{\imath \mathbf{q}\mathbf{x}} \tilde{V}(q).$$

From the total Lagrangian

$$L = L_{el} + L_{phon} + L_{int}$$

one gets according to Euler–Lagrange the equations of motion for the electron:

$$m\ddot{\mathbf{r}}(t) = e\mathbf{E} + \frac{\partial}{\partial \mathbf{r}(t)} \int d\mathbf{x} V(\mathbf{x} - \mathbf{r}(t)) d(\mathbf{x}, t)$$

and

$$\ddot{d}(\mathbf{x}, t) + \omega_0^2 d(\mathbf{x}, t) = -\nabla^2 V(\mathbf{x} - \mathbf{r}(t))$$

for the LO phonons, respectively,

$$\ddot{d}(\mathbf{x}, t) - c^2 \nabla^2 d(\mathbf{x}, t) = -\nabla^2 V(\mathbf{x} - \mathbf{r}(t)),$$

for the LA phonons.

In what follows it is convenient to write the phonon equations in Fourier transforms with the common notation

$$\omega(q) = \begin{cases} \omega_0 & (LO) \\ cq & (LA) \end{cases} \tag{5.1}$$

for the phonon frequencies. We have then

$$\ddot{\tilde{d}}(\mathbf{q}, t) + \omega(q)^2 \tilde{d}(\mathbf{q}, t) = q^2 \tilde{V}(\mathbf{q}) e^{-\imath \mathbf{q}\mathbf{r}(t)}.$$

This equation is easy to solve in favor of the electron coordinate $\mathbf{r}(t)$ with the help of the retarded Green function

$$G_r(t) = \frac{\sin(\omega(q)t)}{\omega(q)}\theta(t).$$

Indeed, if we choose the initial conditions for the phonons as

$$\tilde{d}(\mathbf{q}, 0) = 0, \qquad \dot{\tilde{d}}(\mathbf{q},0) = 0$$

(i.e., at $t = 0$ the phonons are just in rest) we get

$$\tilde{d}(\mathbf{q}, t) = q^2 \int_0^t dt' \frac{\sin\left(\omega(q)(t-t')\right)}{\omega(q)} \tilde{V}(\mathbf{q}) e^{-\imath \mathbf{q}\mathbf{r}(t')}.$$

Introducing this into the electron equation of motion we get a closed equation for the electron coordinate

$$m\ddot{\mathbf{r}}(t) = e\mathbf{E} + \frac{1}{(2\pi)^3} \int_0^t dt' \frac{\partial}{\partial \mathbf{r}(t)} \int d\mathbf{q} \frac{\sin\left[\omega(q)(t-t')\right]}{\omega(q)} e^{\imath \mathbf{q}(\mathbf{r}(t)-\mathbf{r}(t'))} q^2 |\tilde{V}(q)|^2.$$

This is a non-local, non-linear, but causal integro-differential equation one never met in the usual frame of mechanics. So, due to the memory effect you cannot restart the calculation after a while with the new initial data to get the correct result. This group property is valid only for the whole electron-phonon system. However, it is to be expected that after a certain lapse of time the memory will be lost.

Now we have to make a definite choice of the potential $\tilde{V}(q)$:

$$\tilde{V}(q) = \begin{cases} \frac{1}{q^2(q^2+q_D^2)} & (LO) \\ q^{\frac{1}{2}} e^{-q/q_D} & (LA) \end{cases} \tag{5.2}$$

for the two cases. Actually we know from Chap. 4 only the behavior of the potential only for $q \to 0$ and our choice satisfies those requirements. In what concerns the behavior at $q \to \infty$ we preferred here to use a smooth Debye cutoff.

It is now meaningful to introduce dimensionless coordinate and time variables by $q_D x \to x$ and $\omega(q_D)t \to t$. Then we remain actually with only two free dimensionless parameters: a dimensionless "field" $\mathscr{E}$ and a dimensionless positive "coupling constant" C.

The electron equation for coaxial motions (along the electric field) looks as

$$\ddot{x}(t) = \mathscr{E} + C \int_0^t dt' K\left(t - t', x(t) - x(t')\right) \tag{5.3}$$

with a kernel $K(t, x)$ that by our choice of the potential may be expressed through elementary functions. This simplifies essentially as well the discussion of the solutions as the numerical calculations.

It is important to emphasize that no approximations or limits were used in the derivation of this equation. The only important ingredient is the classical description.

An analysis of this equation shows that asymptotic motion with constant velocity occurs indeed, but only in a domain limited by a critical field-coupling ratio for an electron in rest at $t = 0$ and for any field below a certain initial velocity $\dot{x}(0)$ of the electron. Otherwise, a breakthrough occurs and after a certain time it is just the accelerated motion of the electron, as it would occur in the absence of interaction with the phonons.

We illustrate here some typical numerical solutions of this equation.

In the LO case with the choice of the parameters $\mathscr{E} = 0.05$ and $C = 1.0$, having a ratio below the critical one, the function $v(t)$ is shown in Fig. 5.1. After an initial acceleration from the state of rest follow some attenuating oscillations and later the velocity tends indeed to the constant velocity predicted by the analysis of possible asymptotic solutions.

However, with the choice of the parameters $\mathscr{E} = 0.3$ and the same $C = 1.0$ (above the corresponding critical field $\mathscr{E}_c = 1.47057$) one gets an asymptotic accelerated motion as it is shown in Fig. 5.2. But not only a strong field can destroy dissipative behavior, but also a high initial velocity of the electron.

Similar behaviors follow also for the case of LA phonons. However, in the LO case it may happen that an oscillatory behavior around the asymptotic velocity persists!

In the absence of the electrical field the motion of the electron started with the initial velocity $v(0) = 10$ will be damped, as it is illustrated in Fig. 5.3.

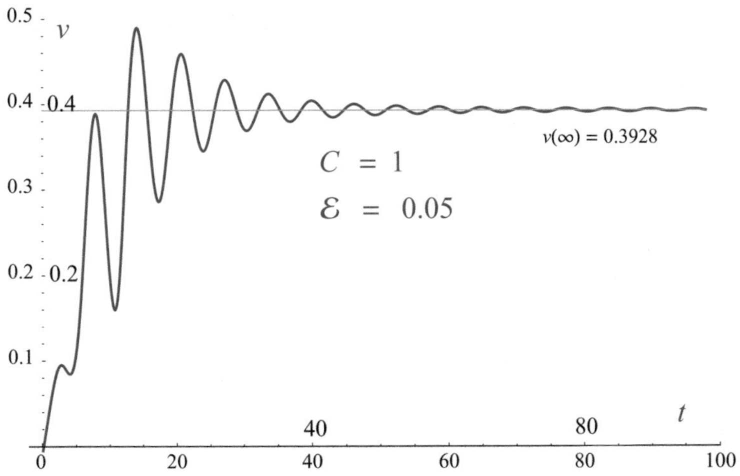

Fig. 5.1 The velocity as function of time for an electric field below the critical one (here $\mathscr{E}/C = 0.05$) by interaction with LO phonons. The horizontal line corresponds to the predicted asymptotic velocity $v(\infty) = 0.392884$

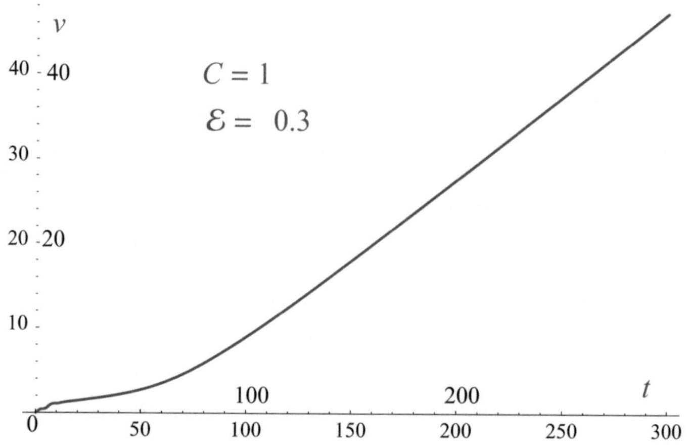

Fig. 5.2 The velocity as function of time for an electric field above the critical one (here $\mathscr{E}/C = 0.3$) by interaction with LO phonons

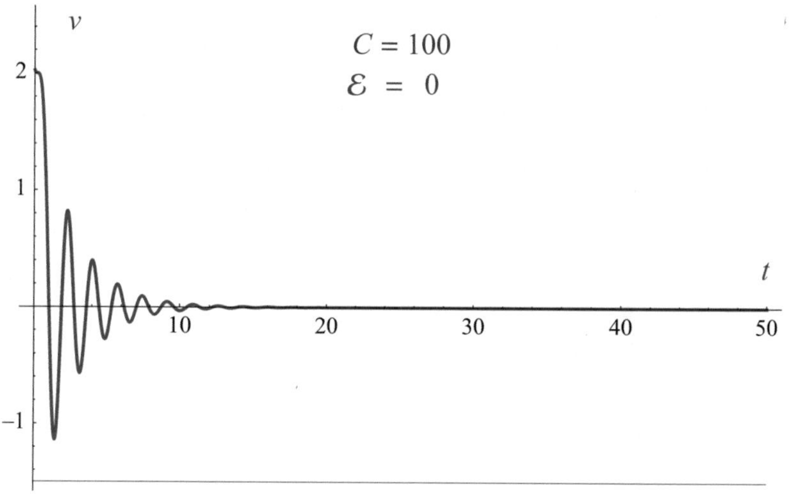

Fig. 5.3 The velocity of the electron starting with $v(0) = 10$ as function of time in the absence of electric field at $C = 100$ by interaction with LA phonons

It is important to remark that also the damping effect occurs only below a critical initial velocity.

Another interesting feature of this model is that the asymptotic stationary velocity in the LA case is proportional to the electric field (ohmic behavior), while by LO it is a non-linear function of the field.

To conclude, the above described solvable model shows that irreversible, dissipative behavior for an open system interacting with a reservoir having a continuous number of degrees of freedom may occur at least in a certain domain of parameters. It is very important to stress this limitation!

Unfortunately, the quantum mechanical description of the same electron-phonon system is not solvable. Therefore, one uses instead certain assumptions about irreversible terms in the equations describing the evolution. While no consistent mathematical proof of those assumptions is available, they proved themselves by very good qualitative and quantitative predictions. In what follows we shall describe these approaches.

5.3 The Boltzmann Equation

In the frame of the classical statistical mechanics a system of point-like particles (inside a Volume Ω) is characterized by the probability density $p(\mathbf{v}, \mathbf{x}, t)$ to find at time instant t a particle with velocity $\mathbf{v}$ at the space-point $\mathbf{x}$, normalized as

$$\int d\mathbf{v} \int_{\Omega} d\mathbf{x} p(\mathbf{v}, \mathbf{x}, t) = 1.$$

The classical Liouville equation for the motion in the presence of electric $\mathbf{E}(\mathbf{x}, t)$ and magnetic fields $\mathbf{B}(\mathbf{x}, t)$ is

$$\left(\frac{\partial}{\partial t} + \mathbf{v} \frac{\partial}{\partial \mathbf{x}} + \frac{1}{m} \mathbf{F}(\mathbf{x}, t) \frac{\partial}{\partial \mathbf{v}} \right) p(\mathbf{k}, \mathbf{x}, t) = 0,$$

where $\mathbf{F} = e(\mathbf{E} + \frac{1}{c} \mathbf{v} \times \mathbf{B})$ is the Lorenz force. The meaning of this equation is that along the trajectory the probability density is conserved. Since the charged particles themselves are sources of fields, one must include their contribution. If the velocity of the particles is very low, one may ignore their contribution to the magnetic field, and for the sake of simplicity we ignore for instance any magnetic field. For the electric field, in a self-consistent manner, one considers the Poisson equation

$$\epsilon_0 \nabla \mathbf{E}(\mathbf{x}, t) = 4\pi \left(\rho(\mathbf{x}, t) + \rho_{ext}(\mathbf{x}, t) \right),$$

where the particle charge density is

$$\rho(\mathbf{x}, t) \equiv eN \left(\int d\mathbf{v} p(\mathbf{v}, \mathbf{x}, t) - \frac{1}{\Omega} \right)$$

and ρ_{ext} is the external charge density. Here we have included also the contribution of a compensating uniform positive charge in the volume Ω that keeps the system stable.

Up to this point we are still in the frame of a reversible mechanical description, however, we may introduce a term describing collisions either on some disordered potential, on phonons in equilibrium, or between the particles themselves, which gives rise to irreversibility.

$$\left(\frac{\partial}{\partial t} + \mathbf{v}\frac{\partial}{\partial \mathbf{x}} + \frac{1}{m}\mathbf{F}(\mathbf{x}, t)\frac{\partial}{\partial \mathbf{v}}\right) p(\mathbf{v}, \mathbf{x}, t) = \frac{\partial p(\mathbf{v}, \mathbf{x}, t)}{\partial t}|_{coll}.$$

In solids, at relatively low particle densities (semiconductors) one considers mostly the scattering on external disturbances (disorder, phonons) and the collision term is obtained from statistical loss-gain considerations

$$\frac{\partial p(\mathbf{v}, \mathbf{x}, t)}{\partial t}|_{coll} = -\int d\mathbf{v}' \left[w(\mathbf{v}, \mathbf{v}')p(\mathbf{v}, \mathbf{x}, t) - w(\mathbf{v}', \mathbf{v})p(\mathbf{v}', \mathbf{x}, t)\right]$$

with transition rates $w(\mathbf{v}, \mathbf{v}')$ characterizing the abrupt velocity changes due to collisions. In solid-state theory, these transition rates are borrowed from the "golden rule" of quantum-mechanics and therefore, we can speak about it as a quasi-classical description. The transition rates are supposed to satisfy the *detailed balance relation* (its justification will be discussed later)

$$w(\mathbf{v}, \mathbf{v}') = e^{\beta(e(\mathbf{v}) - e(\mathbf{v}'))} w(\mathbf{v}', \mathbf{v})$$

implying the kinetic energy $e(\mathbf{v}) = \frac{mv^2}{2}$ before and after the collision. In the peculiar case of elastic scattering, of course, the energy is conserved, while by scattering on phonons it changes due to phonon absorption or emission.

The phonon system at the inverse temperature β acts as a thermal reservoir and (if the external sources are not time dependent and with vanishing boundary conditions!) obviously the stable stationary solution is the Maxwell equilibrium distribution

$$p_0(\mathbf{v}, \mathbf{x}, t) = \frac{1}{\Omega Z} e^{-\beta(\frac{mv^2}{2} + U(\mathbf{x}))},$$

with $U(\mathbf{x})$ being the s.c. equilibrium potential.

These equations conserve the total number of particles, i.e., the normalization

$$\frac{\partial}{\partial t} \int d\mathbf{v} \int_{\Omega} d\mathbf{x} d\mathbf{x} p(\mathbf{v}, \mathbf{x}, t) = 0.$$

In the elastic case, also the total average energy is conserved.

$$\frac{\partial}{\partial t}\left[N \int d\mathbf{v} d\mathbf{x} p(\mathbf{v}, \mathbf{x}, t)\,(e(\mathbf{v}) + U_{ext}(\mathbf{x})) + \frac{1}{2}\int d\mathbf{x} \int d\mathbf{x}' \frac{\rho(\mathbf{x}, t)\rho(\mathbf{x}', t)}{\epsilon_0|\mathbf{x} - \mathbf{x}'|}\right] = 0.$$

However, one may conceive also flow situations by considering appropriate boundary conditions.

5.3.1 Classical Conductivity

Let us now consider a simple choice of the transition rates ("instant relaxation") by

$$w(\mathbf{v}', \mathbf{v}) = \frac{1}{\tau} f_0(e(v)),$$

where τ is a constant "relaxation time" and

$$f_0(e(v)) \equiv \frac{1}{Z} e^{-\beta e(v)}; \quad e(v) = \frac{mv^2}{2}; \quad Z \equiv \int d\mathbf{v} e^{-\beta \frac{mv^2}{2}},$$

which obviously satisfies the balance relation. We shall look now for a spacial homogeneous flow situation in an infinite medium with a constant, homogeneous magnetic field $\mathbf{B}$ and an alternating homogeneous electric field $\mathbf{E}e^{\imath\omega t}$. (Since the uniform charge density of the particles is supposed to be compensated by a background positive charge, in this configuration the external field coincides with the real field in the system!)

In this case for the velocity distribution

$$f(\mathbf{v}, t) = \int d\mathbf{x} p(\mathbf{v}, \mathbf{x}, t)$$

the Boltzmann equation takes a simplified form

$$\left(\frac{\partial}{\partial t} + \frac{e}{m}\left(\mathbf{E}e^{\imath\omega t} + \frac{1}{c}\mathbf{v} \times \mathbf{B}\right)\frac{\partial}{\partial \mathbf{v}}\right) f(\mathbf{v}, t) = -\frac{f(\mathbf{v}, t) - f_0(e(v))}{\tau}.$$

The right-hand side is just an ideal relaxation term with the relaxation time τ. In the stationary flow state, the deviation of the solution from the equilibrium one

$$f(\mathbf{v},t) = f_0(e(v)) + g_1(\mathbf{v})e^{\imath\omega t} + \dots$$

should be at least of first order in the electric field. Retaining only terms of first order in the field we get

$$\left(\imath\omega + \frac{1}{\tau} + \frac{e}{mc}(\mathbf{v} \times \mathbf{B})\frac{\partial}{\partial \mathbf{v}}\right) g_1(\mathbf{v}) = -e\mathbf{E}\mathbf{v}\frac{\partial f_0(e(v))}{\partial e(v)}.$$

To solve this equation, we have to inverse the operator

$$\hat{A} \equiv \imath\omega + \frac{1}{\tau} + \frac{e}{mc}(\mathbf{v} \times \mathbf{B})\frac{\partial}{\partial \mathbf{v}}.$$

This may be easily done by knowing its eigenfunctions and eigenvalues. Let us choose the magnetic field along the z axis $\mathbf{B} = (0, 0, B)$. Then we have the following eigenstates:

$$
\begin{array}{lll}
(\mathbf{e_x v}) & with\ eigenvalue & \imath\omega + \frac{1}{\tau} + \omega_c v_y \\
(\mathbf{e}_y \mathbf{v}) & with\ eigenvalue & \imath\omega + \frac{1}{\tau} - \omega_c v_x \\
(\mathbf{e}_z \mathbf{v}) & with\ eigenvalue & \imath\omega + \frac{1}{\tau},
\end{array}
$$

where $\omega_c = \frac{eB}{mc}$ is the cyclotron frequency. Now one may compute the average electric current density

$$
\langle \mathbf{j} \rangle \equiv e\langle n \rangle \int d\mathbf{v}\mathbf{v} f(\mathbf{v})
$$

(here $\langle n \rangle = \frac{\langle N \rangle}{\Omega}$ is the average particle density) to get the conductivity tensor $\sigma_{\mu\nu}(\omega)$ (defined by $\langle j_\mu \rangle = \sigma_{\mu\nu} E_\nu$) :

$$
\sigma_{zz}(\omega, \omega_c) = \frac{\langle n \rangle e^2}{m} \frac{1}{\imath\omega + \frac{1}{\tau}}
$$

$$
\sigma_{xx}(\omega, \omega_c) = \sigma_{xx}(\omega, \omega_c) = \frac{\langle n \rangle e^2}{2m} \left(\frac{1}{\imath(\omega - \omega_c) + \frac{1}{\tau}} + \frac{1}{\imath(\omega + \omega_c) + \frac{1}{\tau}} \right)
$$

$$
\sigma_{xy}(\omega, \omega_c) = -\sigma_{yx}(\omega, \omega_c) = \frac{\langle n \rangle e^2}{2m} \left(\frac{1}{\imath(\omega - \omega_c) + \frac{1}{\tau}} - \frac{1}{\imath(\omega + \omega_c) + \frac{1}{\tau}} \right)
$$

$$
\sigma_{xz}(\omega, \omega_c) = \sigma_{zx}(\omega, \omega_c) = \sigma_{yz}(\omega, \omega_c) = \sigma_{zy}(\omega, \omega_c) = 0.
$$

Of course, this is an oversimplified model, but nevertheless it shows the role of the relaxation time τ, the phenomenon of cyclotron resonance at $\omega \Rightarrow \omega_c$, as well as the vanishing of some components of the conductivity tensor.

5.4 Kinetic Coefficients

Besides the electrical conductivity there are other kinetic stationary coefficients relating the electrical, as well energy currents to gradients of density or temperature. These non-mechanical coefficients play an important role in semiconductors; however, their derivation is not so simple. The non-mechanical forces cannot be introduced in a microscopical Hamiltonian. We illustrate here the derivation of the Einstein relation between the conductivity and the diffusion coefficient as well as its generalization to obtain other non-mechanical kinetic coefficients.

We define the (stationary) kinetic coefficients $L^{(1)}_{\mu\nu}$ and $L^{(3)}_{\mu\nu}$ as the pure mechanical coefficients, while $\tilde{L}^{(i)}_{\mu\nu}$, $(i = 1, \ldots 4)$ are the non-mechanical ones. Then the phenomenological relationships for the average electric and energy currents are

$$\langle j_\mu \rangle = L^{(1)}_{\mu\nu}\partial_\nu\phi - \tilde{L}^{(1)}_{\mu\nu}T\partial_\nu(\frac{\mu}{T}) - \tilde{L}^{(2)}_{\mu\nu}T\partial_\nu(\frac{1}{T})$$

$$\langle j^E_\mu \rangle = L^{(3)}_{\mu\nu}\partial_\nu\phi - \tilde{L}^{(3)}_{\mu\nu}T\partial_\nu(\frac{\mu}{T}) - \tilde{L}^{(4)}_{\mu\nu}T\partial_\nu(\frac{1}{T}),$$

where ϕ is the electric potential energy. We have chosen as independent macroscopic variables the inverse temperature $\frac{1}{T}$ and the ratio $\frac{\mu}{T}$ of the chemical potential to the temperature.

The coefficients $L^{(1)}_{\mu\nu}$ and $L^{(3)}_{\mu\nu}$ may be calculated either by the classical Boltzmann equation or by Kubo's quantum mechanical theory. Their derivation by linear response theory (in the absence of Coulomb interactions, i.e., within a mean-field approach!) is left for the reader in Chap. 12.

As it is known one might have an equilibrium situation (no currents!), where the gradients of the temperature and chemical potential compensate the current created by the electric potential. Let us assume such a situation occurs and calculate the gradients $\partial_\mu(\frac{\mu}{T})$, $\partial_\mu(\frac{1}{T})$ produced in order to compensate the gradients of the electric potential.

Let us consider equilibrium described by the macro-canonical density matrix

$$e^{-\beta(\mathbf{H}+F-\mu\mathbf{N}-\Omega)}$$

in the presence of the electric potential energy $\phi(\mathbf{x})$ with the term

$$F = \int d\mathbf{x}\,\{\mathbf{n}(\mathbf{x})\phi(\mathbf{x})\}\ ,$$

in the Hamiltonian.

The equilibrium particle density is given by

$$\langle \mathbf{n}(\mathbf{r})\rangle = Tr\{e^{-\beta(\mathbf{H}+F-\mu\mathbf{N}-\Omega)}\mathbf{n}(\mathbf{r})\} \quad .$$

Using the expansion

$$e^{-\beta(\mathbf{H}+F-\mu\mathbf{N})} = e^{-\beta(\mathbf{H}-\mu\mathbf{N})}\left\{1 - \int_0^\beta d\lambda F(-\imath\hbar\lambda) + \ldots\right\},$$

we get

$$\begin{aligned}\langle \mathbf{n}(\mathbf{r})\rangle &= \langle \mathbf{n}(\mathbf{r})\rangle_0 \\ &- \int_0^\beta d\lambda \int d\mathbf{r}' \left(\langle \mathbf{n}(\mathbf{r}', -\imath\hbar\lambda)\mathbf{n}(\mathbf{r})\rangle_0 - \langle \mathbf{n}(\mathbf{r}', -\imath\hbar\lambda)\rangle_0\langle \mathbf{n}(\mathbf{r})\rangle_0\right)\phi(\mathbf{r}'),\end{aligned}$$

where the symbol $\langle\ldots\rangle$ means $\frac{Tr\{e^{-\beta(\mathbf{H}-\mu\mathbf{N})}\ldots\}}{Tr\{e^{-\beta(\mathbf{H}-\mu\mathbf{N})}\}}$.

Taking into account the macroscopic homogeneity in the absence of the field, by a partial integration we get

$$\partial_\nu \langle n(\mathbf{r})\rangle =$$
$$-\int_0^\beta d\lambda \int d\mathbf{r}' \partial_\nu \left(\langle \mathbf{n}(\mathbf{r}',-\imath\hbar\lambda)\mathbf{n}(\mathbf{r})\rangle_0 - \langle \mathbf{n}(\mathbf{r}',-\imath\hbar\lambda)\rangle_0 \langle \mathbf{n}(\mathbf{r})\rangle_0 \right) \phi(\mathbf{r}')$$
$$= -\int_0^\beta d\lambda \int d\mathbf{r}' \left(\langle n(\mathbf{r}',-\imath\hbar\lambda) n(\mathbf{r})\rangle_0 - \langle n(0,-\imath\hbar\lambda)\rangle_0 \langle n(0)\rangle_0 \right) \partial_\nu \phi(\mathbf{r}').$$

(Here in neglecting the surface integral terms, the vanishing of long range correlations was admitted.) If the field does not vary too rapidly with the coordinate, taking into account that the operator of the total particle number commutes with the Hamiltonian, we have further

$$\partial_\nu \langle \mathbf{n}(\mathbf{r}\rangle = -\frac{1}{VkT} \left(\langle \mathbf{N}^2\rangle_0 - \langle \mathbf{N}\rangle_0^2 \right) \partial_\nu \phi(\mathbf{r}).$$

In the same way one may get also

$$\partial_\nu \langle \mathbf{h}(\mathbf{r})\rangle = -\frac{1}{VkT} \left(\langle \mathbf{HN}\rangle_0 - \langle \mathbf{H}\rangle_0 \langle \mathbf{N}\rangle_0 \right) \partial_\nu \phi(\mathbf{r}).$$

However, as it may be seen easily that

$$\left(\langle \mathbf{N}^2\rangle_0 - \langle \mathbf{N}\rangle_0^2 \right) = \frac{\partial \langle \mathbf{N}\rangle_0}{\partial \beta\mu} = V \frac{\partial \langle \mathbf{n}(\mathbf{r})\rangle_0}{\partial \beta\mu}$$

and

$$\left(\langle \mathbf{NH}\rangle_0 - \langle \mathbf{N}\rangle_0 \langle \mathbf{H}\rangle_0 \right) = \frac{\partial \langle \mathbf{H}\rangle_0}{\partial \beta\mu} = V \frac{\partial \langle \mathbf{h}(\mathbf{r})\rangle_0}{\partial \beta\mu}.$$

Therefore, from the above equations it follows:

$$\partial_\nu \left(\frac{\mu}{T} \right) = -\left[\frac{\partial\left(\frac{\mu}{T}\right)}{\partial \langle n\rangle} \frac{\partial \langle n\rangle}{\partial\left(\frac{\mu}{T}\right)} + \frac{\partial\left(\frac{\mu}{T}\right)}{\partial \langle h\rangle} \frac{\partial \langle h\rangle}{\partial\left(\frac{\mu}{T}\right)} \right] \frac{1}{T} \partial_\nu \phi$$
$$= -\frac{d\left(\frac{\mu}{T}\right)}{d\left(\frac{\mu}{T}\right)} \frac{e}{T} \partial_\nu \phi = -\frac{1}{T} \partial_\nu \phi,$$

while

$$\partial_\nu\left(\frac{1}{T}\right) = -\left[\frac{\partial\left(\frac{1}{T}\right)}{\partial\langle n\rangle}\frac{\partial\langle n\rangle}{\partial\left(\frac{\mu}{T}\right)} + \frac{\partial\left(\frac{1}{T}\right)}{\partial\langle h\rangle}\frac{\partial\langle h\rangle}{\partial\left(\frac{1}{T}\right)}\right]\frac{e}{T}\partial_\nu\phi = -\frac{d\left(\frac{1}{T}\right)}{d\left(\frac{\mu}{T}\right)}\frac{1}{T}\partial_\nu\phi = 0.$$

From the conditions $\langle j_\mu\rangle = \langle j_\mu^E\rangle = 0$, together with the above relations we get immediately

$$\tilde{L}_{\mu\nu}^{(1)} = L_{\mu\nu}^{(1)}, \qquad \tilde{L}_{\mu\nu}^{(3)} = L_{\mu\nu}^{(3)}.$$

The first of these relations is just the well-known Einstein relation between the electrical conductivity and the diffusion coefficient.

Now we may rewrite the general expressions of the average currents as

$$\langle j_\mu\rangle = L_{\mu\nu}^{(1)}[(\partial_\nu\phi - T\partial_\nu(\frac{\mu}{T})] - \tilde{L}_{\mu\nu}^{(2)}T\partial_\nu(\frac{1}{T})$$

$$\langle j_\mu^E\rangle = L_{\mu\nu}^{(3)}[\partial_\nu\phi - T\partial_\nu(\frac{\mu}{T})] - \tilde{L}_{\mu\nu}^{(4)}T\partial_\nu(\frac{1}{T}).$$

To derive the expressions of the remaining coefficients $\tilde{L}_{\mu\nu}^{(2)}$ and $\tilde{L}_{\mu\nu}^{(4)}$ one has to develop a little bit of fantasy. The Luttinger approach follows the idea of the Einstein relation, by introducing the interaction with a fictitious gravitational field $\psi(\mathbf{x})$ interacting with the Hamiltonian density $\rho^E(\mathbf{x})$ as

$$F' = \int d\mathbf{x}\rho^E(\mathbf{x})\psi(\mathbf{x})$$

and again requiring a compensation of all currents in equilibrium.

The final table of the quantum mechanical kinetic coefficients looks then as

$$L_{\mu\nu}^{(1)} = \lim_{s\to+0}\lim_{\Omega\to\infty}\Omega\int_0^\infty dte^{-st}\int_0^\beta d\lambda\langle\mathbf{j}_\nu(t-\imath\hbar\lambda)\mathbf{j}_\mu(0)\rangle_0, \tag{5.4}$$

$$\tilde{L}_{\mu\nu}^{(2)} = \lim_{s\to+0}\lim_{\Omega\to\infty}\Omega\int_0^\infty dte^{-st}\int_0^\beta d\lambda\langle\mathbf{j}_\nu^E(t-\imath\hbar\lambda)\mathbf{j}_\mu(0)\rangle_0,$$

$$L_{\mu\nu}^{(3)} = \lim_{s\to+0}\lim_{\Omega\to\infty}\Omega\int_0^\infty dte^{-st}\int_0^\beta d\lambda\langle\mathbf{j}_\nu(t-\imath\hbar\lambda)\mathbf{j}_\mu^E(0)\rangle_0,$$

$$\tilde{L}_{\mu\nu}^{(4)} = \lim_{s\to+0}\lim_{\Omega\to\infty}\Omega\int_0^\infty dte^{-st}\int_0^\beta d\lambda\langle\mathbf{j}_\nu^E(t-\imath\hbar\lambda)\mathbf{j}_\mu^E(0)\rangle_0.$$

We have to stress that neither of these relationships is valid in the presence of the Coulomb interactions. Also they may be used only within the mean-field approximation. Another difficulty lies in the problematic definition of the energy current density operators in the presence of interaction with phonons.

5.5 Master and Rate Equations

5.5.1 Master Equations

Obviously, a quasi-classical description as in the previous subsection is of very limited validity. From a quantum mechanical point of view, simultaneous coordinate and momentum are not the proper characterization of a particle. Actually, some quantum numbers characterize the state. Therefore within the same statistical gain-loss philosophy one uses the so-called Master and rate equations.
The statistical state ("mixed state") of a quantum mechanical system may be defined by the probability $P_n(t)$ to find the system at instant t in the eigenstate $|n\rangle$ of an "unperturbed" Hamilton operator H_0. The time-evolution of this statistical system is due to some interaction H' causing transitions between the unperturbed states $|n\rangle$ and one assumes that the transition rates $W_{nn'}$ are given by the "golden rule" of adiabatic perturbation theory:

$$W_{nn'} \equiv \lim_{\Delta t \to \infty} \frac{1}{\Delta t} \left|\langle n'|U(\Delta t)|n\rangle\right|^2$$

with $U(\Delta t)$ being the unitary evolution operator during the time interval Δt.
The Master equation looks then as

$$\frac{\partial}{\partial t} P_n(t) = -\sum_{n'} (P_n(t)W_{nn'} - P_{n'}(t)W_{n'n}) \equiv -\sum_{n'} A_{nn'} P_{n'}(t).$$

The general properties of the transitions rates emerge from the "golden rule" definition. The transition rates are obviously non-negative and satisfy the generalized balance property emerging from the unitarity of U:

$$W_{nn'} \geq 0; \qquad \sum_{n'} (W_{nn'} - W_{n'n}) = 0.$$

In virtue of these properties for the solution of the Master equation it follows that:

(i) normalization is conserved $\frac{\partial}{\partial t} \sum_{n'} P_{n'}(t) = 0$,
(ii) positivity is conserved, i.e., $P_n(t) \geq 0$ for any $t > 0$ if it $P_n(0) \geq 0$. (it emerges from the fact that $P_n(t) = 0$ implies $\frac{\partial}{\partial t} P_n(t) \geq 0$).
(iii) the energy is conserved due to the admitted adiabatic coupling–decoupling of the interaction,
(iv) the evolution leads to a stable equilibrium state (irreversibility).

The formal solution is

$$P(t) = e^{-At} P(0)$$

and the operator A has non-negative eigenvalues.

Indeed, if $\xi_n = x_n + \imath y_n$ is a complex number, then in thermal equilibrium.

$$\Re\left\{\sum_{n,n'} \xi_n^* A_{nn'}\xi_{n'}\right\} = \Re\left\{\sum_{n,n'} W_{nn'}\xi_n^* \left(\xi_n - \xi_{n'}\right)\right\}$$
$$= \frac{1}{2}\sum_{n,n'} W_{n'n}\left[(x_n - x_{n'})^2 + (y_n - y_{n'})^2\right] \geq 0,$$

where we used also the balance property. If a general connectivity occurs, i.e., any state may be achieved from any state through successive transitions, then there is a single eigenstate with eigenvalue 0 and its components are all equal ($\xi_n^0 = const.$). It follows therefore

$$\lim_{t\to\infty} P(t) = P^0,$$

where $P_n^0 = const.$ and

$$P(t) = e^{-At}(P(0) - P^0) + P^0.$$

Such an evolution is called a Markowian process.

Let us now assume that we have a system in contact with a macroscopic thermostat, whose state may be admitted remaining unchanged in thermal equilibrium Then the probability of the total state (system plus thermostat) may be admitted to be

$$P_{n\alpha}(t) = \mathscr{P}_n(t)\mathscr{P}_\alpha^0; \qquad \mathscr{P}_\alpha^0 = \frac{e^{-\beta E_\alpha}}{\sum_{\alpha'} e^{-\beta E_{\alpha'}}},$$

where the quantum numbers α characterize the states of the thermostat.
From the Master equation for $P_{n\alpha}(t)$ it follows:

$$\frac{\partial}{\partial t}\mathscr{P}(t) = -\mathscr{A}\,\mathscr{P}(t); \qquad \mathscr{A}_{nn'} = \delta_{nn'}\sum_{n''}\mathscr{W}_{nn'} - \mathscr{W}_{n'n}.$$

From the energy conservation ($E_n + E_\alpha = E_{n'} + E_{\alpha'}$) and the balance property of the total system follows the balance property of the averaged transition rates $\mathscr{W}_{nn'}$

$$\mathscr{W}_{nn'} \equiv \sum_{\alpha\alpha'} \mathscr{P}_\alpha^0 W_{n\alpha,n'\alpha'}$$

$$\sum_{n'}\left(\mathscr{W}_{nn'} - e^{\beta(E_n - E_{n'})}\mathscr{W}_{n'n}\right) = 0.$$

The operator $\mathscr{A}$ again has (under a similar connectivity condition) only eigenvalues with non-negative real part, whereas the only eigenstate corresponding to the zero eigenvalue is

$$P_n^0 = \frac{e^{-\beta E_n}}{\sum_{n'} e^{-\beta E_{n'}}}.$$

5.5.2 *Rate Equations*

Let us consider the case, where the index $n \equiv \{n_1, n_2, \ldots)$ corresponds to the set of occupation numbers of some one-electron states $|i>$, $(i = 1, 2, \ldots)$. The energy (without any interaction between the electrons!) is

$$E_n = \sum_i e_i n_i$$

and let us consider that through interaction with some thermal bath (for example, phonons) transitions occur. In the lowest order of adiabatic perturbation theory the transition rate $\mathscr{W}_{nn'}$ is the sum of one-electron transition rates

$$\mathscr{W}_{nn'} = \sum_{i,j} w_{ij} n_i (1 - n'_j) \Pi_{l \neq i,j} \delta_{n_l, n'_l}.$$

Within this lowest order perturbation approximation even a "detailed balance relation"

$$w_{ij} = w_{ji} e^{\beta(e_i - e_j)}$$

occurs. Then the evolution of the average occupation number $\langle n_i(t) \rangle$ of a state i

$$\langle n_i(t) \rangle \equiv \sum_n n_i P_n(t)$$

looks as

$$\frac{\partial}{\partial t} \langle n_i(t) \rangle = -\sum_{n,n'} W_{nn'} (n_i - n'_i) P_n(t) = -\sum_{i,j} \left[w_{ij} \langle n_i (1 - n_j) \rangle - w_{ji} \langle n_j (1 - n_i) \rangle \right].$$

As we see the equation is not closed, it includes the next average $\langle n_i n_j \rangle$. However, if one admits the approximation

$$\langle n_i n_j \rangle \approx \langle n_i \rangle \langle n_j \rangle$$

one gets a closed non-linear rate equation for the fermions

$$\frac{\partial}{\partial t}\langle n_i(t)\rangle = -\sum_{i,j}\left[w_{ij}\langle n_i(t)\rangle\left(1-\langle n_j(t)\rangle\right) - w_{ji}\langle n_j(t)\rangle\left(1-\langle n_i(t)\rangle\right)\right].$$

Here obviously the Pauli principle is respected only on average. The average occupation numbers remain in the interval [0, 1] if they were in this interval and under the previously discussed connectivity condition the stable asymptotic solution is the Fermi function

$$f(\epsilon_i) = \frac{1}{e^{\beta(\epsilon_i-\mu)}+1}.$$

The new constant μ here is the chemical potential fixed by the average total number of the fermions, which is conserved by the transitions.

5.6 Hopping Transport

A peculiar transport mechanism may occur in solids, when conduction occurs through jumping between localized states. This is a typical situation in disordered semiconductors or, as we shall see later, even in crystalline semiconductors under ultra-strong magnetic fields.

Let us consider here a system of neutral fermions on localized states on an arbitrary disordered or ordered lattice points $\mathbf{r}_i$. The state of the fermion system is therefore described by the set of occupation numbers $n_i = 0, 1$ of the localized states having the one-particle energies ϵ_i. The total energy is

$$E_n = \sum_i \epsilon_i n_i.$$

We shall describe the time-evolution of this system again by a rate equation.

The transition rates w_{ij} are assumed to obey the balance property of the preceding subsection. The only global observable we may consider beside the total energy is the average particle density

$$\langle n(\mathbf{x},t)\rangle = \sum_i \langle n_i(t)\rangle \delta(\mathbf{x}-\mathbf{x}_i).$$

In the case of charged particles (electrons) the particle density would define also the electric potential created by them.

Let us assume that we want to describe the evolution of a state which is not far away from equilibrium described by the Fermi function $f_i \equiv f(\epsilon_i)$, i.e.,

$$\langle n_i(t)\rangle = f_i + \delta\langle n_i(t)\rangle$$

with $\delta\langle n_i(t)\rangle$ very small. Then we may linearize the rate equation as

$$\frac{\partial}{\partial t}\delta\langle n(t)\rangle = -\Gamma\delta\langle n(t)\rangle$$

with

$$\Gamma_{ij} \equiv \left(\delta_{ij}\sum_l \mathscr{W}_{il} - \mathscr{W}_{ij}\right)\frac{1}{f_i(1-f_i)}$$

and

$$\mathscr{W}_{ij} = f_i(1-f_j)w_{ij} = \mathscr{W}_{ji}.$$

The real matrix Γ is Hermitian in the scalar product

$$(\psi, \phi) \equiv \sum_i \psi_i^* \phi_i \frac{1}{f_i(1-f_i)}$$

and non-negative since

$$(\psi, \Gamma\psi) = \frac{1}{2}\sum_{i,j}\left|\frac{\psi_i}{f_i(1-f_i)} - \frac{\psi_j}{f_j(1-f_j)}\right|^2 \mathscr{W}_{ij} \geq 0.$$

Assuming connectivity of the lattice, there is just one null-eigenvector

$$\xi_i = \frac{f_i(1-f_i)}{\sum_j \sqrt{f_j(1-f_j)}}.$$

Therefore, the deviation from the equilibrium solution

$$\delta\langle n(t)\rangle = e^{-\Gamma t}\delta\langle n(0)\rangle$$

always vanishes asymptotically, $\delta\langle n(t)\rangle \to 0$ as $t \to \infty$. The null-eigenvector plays no role here, since the total particle number conservation implies

$$\sum_i \delta\langle n(t)\rangle = 0$$

i.e., $\delta\langle n(t)\rangle$ has no components in the direction of ξ

$$(\xi, \delta\langle n(t)\rangle) = 0.$$

5.6.1 Hopping Diffusion on a Periodic Cubic Lattice

One can find a simple explicit solution for the hopping on a periodical cubic Bravais lattice of states of equal energies with translation invariant transition rates $w(\mathbf{R}, \mathbf{R}') \equiv w(\mathbf{R} - \mathbf{R}') > 0$ of finite range. The detailed balance here looks as

$$w(\mathbf{R} - \mathbf{R}') = w(\mathbf{R}' - \mathbf{R}).$$

The discrete Fourier transformation $\delta\tilde{n}(\mathbf{k}) \equiv \sum_{\mathbf{R}} \delta\langle n_{\mathbf{R}}\rangle e^{\imath\mathbf{kR}}$ is helpful here and we get the simple scalar equation

$$\frac{\partial}{\partial t}\delta\tilde{n}(\mathbf{k}, t) = -\tilde{\Gamma}(\mathbf{k})\delta\tilde{n}(\mathbf{k}, t)$$

with

$$\tilde{\Gamma}(\mathbf{k}) = \tilde{w}(0) - \tilde{w}(\mathbf{k}).$$

From the reality of Γ it follows $\tilde{\Gamma}(\mathbf{k}) = \tilde{\Gamma}(-\mathbf{k})^*$,while from the detailed balance follows that $\tilde{\Gamma}(\mathbf{k})$ is real. On the other hand, the inequality

$$\sum_{\mathbf{R}} w(\mathbf{R})e^{\imath\mathbf{kR}} \leq \sum_{\mathbf{R}} |w(\mathbf{R})e^{\imath\mathbf{kR}}|$$

implies $\tilde{\Gamma}(\mathbf{k}) > 0$ for $\mathbf{k} \neq 0$. The total particle number conservation in Fourier transforms looks as $\delta\tilde{n}(0) = 0$. The assumed finite range of the transition rates implies that one may develop $\tilde{\Gamma}(\mathbf{k})$ around $\mathbf{k} = 0$ and the lowest term is a quadratic one $\tilde{\Gamma}(\mathbf{k}) \approx Dk^2$, with

$$D = \frac{1}{6}\sum_{\mathbf{R}} w(\mathbf{R})\mathbf{R}^2 > 0.$$

Asymptotically for $t \to \infty$ the behavior of the particle density is dominated by the small k values of the Fourier transform for which

$$\delta\tilde{n}(\mathbf{k}, \mathbf{t}) \approx e^{-Dk^2 t}\delta\tilde{n}(\mathbf{k}, 0)$$

and this corresponds exactly to the behavior predicted by the diffusion equation

$$\frac{\partial}{\partial t}n(\mathbf{x}, t) = -D\nabla^2 n(\mathbf{x},t).$$

Therefore, we may identify the constant D with the diffusion constant of this hopping model.

5.6.2 *Transverse Magneto-Resistance in Ultra-Strong Magnetic Field*

As we have earlier discussed in Sect. 2.2.2, the Landau states of an electron in a constant magnetic field are characterized by a discrete quantum number n, a wave vector k_z in the field direction, and the coordinate X of the center of cyclotron motion in the plane perpendicular to the magnetic field. The extension of the wave-function in the x direction is determined by the magnetic length $\ell_B = \sqrt{\frac{\hbar c}{|e|B}}$, therefore in a very strong magnetic field $\mathbf{B}$ one might expect that at least along the x axis the state is strongly localized, and the rapidly, with the frequency $\omega_c = \frac{|e|B}{mc}$, oscillating relative coordinate is irrelevant for transport phenomena. In this sense, one might conceive the transport in the x direction as a hopping problem in continuum from one X to another one X' due to the interaction with phonons or lattice defects.

In the presence of an electric field $\mathbf{E} = (E, 0, 0)$ in the x direction the degeneracy of the Landau state is lifted. Intuitively one may expect a potential energy contribution

$$\epsilon_{n,X,k_z,} = \frac{\hbar^2 k_z^2}{2m} + \hbar\omega_c(n + \frac{1}{2}) + eEX \qquad (n = 0, 1, 2, \ldots).$$

(More precisely, one may show that there is also a constant non-relevant energy shift $-\frac{e^2E^2}{m\omega_c}$, that we may ignore.)

However, if hopping between the states of different coordinates X is allowed, the contribution of the field produced by the non-homogeneous distribution of the electron charge modifies in time the effective on-site field, and finally, a self-consistent in-homogeneous equilibrium distribution arises which screens the electric field inside far away from the surface. We shall not discuss here this kind of evolution, but consider another situation in which we admit that a steady flow state with constant electric field is achieved by attaching some external electron sources at the ends of the system along the x axis.

The average X coordinate is

$$\langle X \rangle \equiv \sum_\nu X \langle n_\nu \rangle$$

with the simplifying notation $\nu \equiv n, k_z, X$.
The average velocity by hopping (omitting the contribution of the contacts on the boundaries!) is

$$\langle \dot{X} \rangle = \sum_\nu X \frac{d}{dt} \langle n_\nu \rangle = -\sum_{\nu,\nu'} X \left\{ W_{\nu\nu'} \langle n_\nu \rangle (1 - \langle n_{\nu'} \rangle) - W_{\nu'\nu} \langle n_{\nu'} \rangle (1 - \langle n_\nu \rangle) \right\},$$

where the transition rates due to phonons or random potentials (impurities) satisfy the detailed balance

$$W_{\nu\nu'} = W_{\nu'\nu} e^{\beta(\epsilon_\nu - \epsilon_{\nu'})}.$$

We assume now that a steady flow state occurs in which the charge density is unchanged, i.e., the average occupation numbers are the same as in equilibrium in the absence of the electric field

$$\langle n_{n,k_z,X}\rangle = f(\epsilon^0_{n,k_z}); \qquad \epsilon^0_{n,k_z} = \frac{\hbar^2 k_z^2}{2m} + \hbar\omega_c(n + \frac{1}{2})$$

with $f(\epsilon)$ being the Fermi function. (Later we shall check the consistency of this assumption.)
Introducing a symmetrical matrix $\tilde{W}_{\nu\nu'}$ by

$$W_{\nu\nu'} = \tilde{W}_{\nu\nu'} e^{\frac{\beta}{2}(E_\nu - E_{\nu'})}; \qquad \tilde{W}_{\nu\nu'} = \tilde{W}_{\nu'\nu}$$

we may then write

$$\langle \dot{X}\rangle = -\sum_{\nu,\nu'} X\tilde{W}_{\nu\nu'} \left\{ e^{\frac{\beta}{2}(E_\nu - E_{\nu'})} f(\epsilon^0_{n,k_z}) \left(1 - f(\epsilon^0_{n',k_z'})\right) \right.$$
$$\left. - e^{-\frac{\beta}{2}(E_\nu - E_{\nu'})} f(\epsilon^0_{n',k_{z'}}) \left(1 - f(\epsilon^0_{n,k_z})\right)\right\}.$$

Now, if one is interested only in the electric conductivity one may retain only the first order terms in the electric field. Since the terms inside the big bracket are vanishing in the absence of the field we may replace $\tilde{W}_{\nu\nu'}$ by its equilibrium value $w_{\alpha,X;\alpha',X'}$ in the absence of the electric field (here we introduced the notation α for the subset of quantum numbers n, k_z) and retain only the term of first order in the field from the exponentials in the big bracket

$$\langle \dot{X}\rangle \approx -e\beta E \sum_{\alpha,X}\sum_{\alpha',X'} X(X - X')w_{\alpha,X;\alpha',X'} f(\epsilon^0_{n,k_z}) \left(1 - f(\epsilon^0_{n',k_z'})\right)$$
$$= \; -\tfrac{1}{2}e\beta E \sum_{\alpha,X}\sum_{\alpha',X'} (X - X')^2 w_{\alpha,X;\alpha',X'} f(\epsilon^0_\alpha) \left(1 - f(\epsilon^0_{\alpha'})\right).$$

Then we get the current density along the x axis as

$$j_x \equiv \frac{e}{\Omega}\langle \dot{X}\rangle = -\frac{1}{2\Omega} e^2\beta E \sum_{\alpha,X}\sum_{\alpha',X'} (X - X')^2 w_{\alpha,X;\alpha',X'} f(\epsilon^0_\alpha) \left(1 - f(\epsilon^0_{\alpha'})\right)$$

and we may identify the transverse conductivity (Titeica formula)

$$\sigma_{xx} = \frac{1}{2\Omega} e^2 \beta E \sum_{\alpha,X} \sum_{\alpha',X'} (X - X')^2 w_{\alpha,X;\alpha',X'} f(\epsilon_\alpha^0) \left(1 - f(\epsilon_{\alpha'}^0)\right),$$

where Ω is the volume of the system. Of course, as usual the thermodynamic limit has still to be performed.

Let us now check the validity of our assumption about the stationary flow solution. That means $f(\epsilon_\alpha^0)$ should be a stationary solution of the rate equation, at least to first order in the electric field. Under the same kind of expansion, we get the requirement

$$\sum_{\alpha',X'} (X - X') w_{\alpha,X;\alpha',X'} f(\epsilon_\alpha^0) \left(1 - f(\epsilon_{\alpha'}^0)\right) = 0.$$

If the transition rates depend only on the distance $|X - X'|$ and fall rapidly with it, then far away from the boundaries this condition is obviously satisfied. At the boundaries one might suppose the presence of some external sources to feed the current flow.

5.6.3 Seebeck Coefficient for Hopping Conduction on Random Localized States

We want to derive now the Seebeck coefficient for hopping conduction on randomly distributed localized states i of coordinate $\mathbf{x}_i$ and energy ϵ_i. We have in mind mainly amorphous semiconductors, where (in the average) a continuous energy spectrum of localized states in the so-called mobility gap replaces the forbidden gap of the crystal. We shall not use the Luttinger formulas of Sect. 5.4 for the kinetic coefficients, but the assumption of particle and energy flow in a stationary local equilibrium distribution state with coordinate dependent temperature $T(\mathbf{x})$ and chemical potential $\mu(\mathbf{x})$ in the x-direction. We suppose that on the average the system is homogeneous and isotropic. The average electron occupation of the localized state is then

$$\langle n_i \rangle = \frac{1}{(e^{\epsilon_i} - \mu_i)/k_B T_i + 1},$$

where

$$\frac{1}{T_i} = \frac{1}{T} + (x_i - \bar{x}) \nabla_x \left(\frac{1}{T}\right)$$

and

$$\frac{\mu_i}{T_i} = \frac{\mu}{T} + (x_i - \bar{x})\nabla_x\left(\frac{\mu}{T}\right).$$

Then for small gradients we get

$$\langle n_i \rangle \approx f_i - \frac{1}{k_B} f_i(1 - f_i)(x - \bar{x})\left[\nabla_x\left(\frac{\mu}{T}\right) - \epsilon_i \nabla_x\left(\frac{1}{T}\right)\right].$$

On the other hand the average charge flow per unit surface across a plane surface S perpendicular to the x axis at $x = x_s$ is given by

$$J_x^S(x_s) = \frac{e}{S}\sum_{i,j}\theta(x_i - x_s)\theta(x_s - x_j)\left[\langle n_i\rangle(1-\langle n_j\rangle)W_{ij}(x_i - x_j) - \langle n_j\rangle(1-\langle n_i\rangle)W_{ji}\right],$$

or taking its average over the position x_s of the plane

$$\bar{J}_x = \frac{1}{L}\int_0^L dx_s J_x^S = \frac{e}{\Omega}\sum_{i,j}\langle n_i\rangle(1 - \langle n_j\rangle)W_{ij}(x_i - x_j).$$

Inserting the above expression for the average occupation number $\langle n_i \rangle$ and taking into account that $f_i(1 - f_j)W_{ij}(x_i - x_j)$ either for tunneling or for phonon-assisted hopping is anti-symmetric due to the detailed balance, we get

$$\begin{aligned}\bar{J}_x &= \frac{e}{\Omega}\sum_{i,j} W_{ij}(x_i - x_j)\delta\langle n_i\rangle \\ &= \frac{e}{k_B\Omega}\sum_{i,j} f_i(1 - f_j)W_{ij}(x_i - x_j)(x_i - \bar{x})\left[\nabla_x\left(\frac{\mu}{T}\right) - \epsilon_i\nabla_x\left(\frac{1}{T}\right)\right].\end{aligned}$$

Then, choosing the arbitrary coordinate $\bar{x}$ conveniently as

$$\bar{x} = \frac{\sum_{i,j} W_{ij}(x_i - x_j)(\epsilon_i - \epsilon_j)(x_i + x_j)/2}{\sum_{i,j} W_{ij}(x_i - x_j)(\epsilon_i - \epsilon_j)}$$

we get

$$\bar{J}_x = \frac{\sigma T}{e}\nabla_x\left(\frac{\mu}{T}\right) + \chi T\nabla_x\left(\frac{1}{T}\right)$$

with

$$\chi = -\frac{e}{2\Omega K_B T}\sum_{i,j} f_i(1 - f_j)W_{ij}(x_i - x_j)^2(\epsilon_i + \epsilon_j)/2,$$

while the conductivity

$$\sigma = \frac{e^2}{2\Omega k_B T} \sum_{i,j} f_i(1 - f_j) W_{ij} (x_i - x_j)^2$$

follows from the Einstein relation.

$$\langle j_x \rangle = -\frac{\sigma}{e} \left[\nabla_x \phi - T \nabla_x \left(\frac{\mu}{T} \right) \right] + \chi T \nabla_x \left(\frac{1}{T} \right) \tag{5.5}$$

we already discussed in Sect. 5.4. For identification of notations with the previous ones: $L_{11}^{(1)} = -\sigma/e$.

The thermoelectric force is defined as the ratio of the created electric potential to the imposed temperature gradient in the equilibrium state

$$\alpha = \frac{\nabla(\phi - \mu)}{e \nabla T}; \qquad (\langle j_x \rangle = 0). \tag{5.6}$$

Using the above definitions of the kinetic coefficients α given by

$$\alpha = \frac{1}{eT} (\frac{e\chi}{\sigma} - \mu) = \frac{k_B}{e} S$$

with the dimensionless Seebeck coefficient

$$S = \frac{1}{k_B T} (\frac{e\chi}{\sigma} - \mu). \tag{5.7}$$

Let us define

$$\mathscr{P}_{ij} = f_i(1 - f_j) W_i j (x_i - x_j)^2. \tag{5.8}$$

Then with the coefficients we found for hopping we have

$$S_\mu = \frac{1}{k_B T} \frac{\sum_{i,j} \mathscr{P}_{ij} (\epsilon_i + \epsilon_j - 2\mu)/2}{\sum_{i,j} \mathscr{P}_{ij}}.$$

For very low temperatures $\mathscr{P}_{ij}$ vanishes very rapidly outside the domain

$$|e_i - \mu| / k_B T \ll 1; \qquad |e_j - \mu| / k_B \ll 1.$$

In the case of tunneling the initial and final energies are just the same, but even by phonon assisted tunneling at low temperatures these two energies must be very close to each other. Therefore one may expect that the Seebeck coefficient S_μ at low temperature goes to zero.

However, the picture changes essentially if one takes into account that there is a repulsion U in the double occupation states. The equilibrium distribution, as it was shown in Sect. 2.5.4 in this case is not the Fermi function. For a very large repulsion U/k_BT and motion of electrons on simple occupied states one may approximate the new equilibrium distribution function by a Fermi function with a shifted chemical potential $\mu \rightarrow \mu + k_BT \ln 2$. On the other hand, the relation between the kinetic coefficients remains the same, i.e., Eqs. 5.5–5.7 remain unchanged, but the energies in $\mathscr{P}_{ij}$ now are around $\mu* = \mu + k_BT \ln 2$. The Seebeck coefficient is thus modified by a shift

$$S_\mu \rightarrow S_{\mu*} - \ln 2, \qquad (\mu* = \mu + k_BT). \tag{5.9}$$

Another extreme situation may occur in the hole transport, i.e., transitions between simple and double occupied states, in the extreme fall of large repulsion the situation is described with a Fermi function with a shifted chemical potential $\mu \rightarrow \mu - k_BT \ln 2$. Therefore the Seebeck coefficient is modified by a shift

$$S_\mu \rightarrow S_{\mu*} + \ln 2, \qquad (\mu* = \mu - k_BT). \tag{5.10}$$

Since, as we mentioned, S_μ or $S_{\mu*}$ is generally speaking very small at low temperatures, taking into account also other situations as the two discussed extreme cases, generally speaking one would expect a low temperature plateau of the Seebeck coefficient in amorphous semiconductors

$$S \rightarrow \xi \ln 2; \qquad (|\xi| \leq 1).$$

It is important to remark that the energy transport in our treatment, as well as in other approaches, like that in the frame of the Boltzmann equation, is reduced to the energy transport by the electrons.

Chapter 6
Optical Properties

The linear response to a time-dependent adiabatic external perturbation as well as the equilibrium linear response are presented. The specific relationships of the longitudinal dielectric response of an electron–hole plasma to the density-density correlation function depending on the explicit inclusion of the Coulomb interaction between the charged particles are explained. The dielectric function is explicitly calculated within the Hartree approximation. The response to a transverse (propagating) electric field within a simple model of a semiconductor with s-type conduction and p-type valence bands including the Coulomb attraction between the created electron–hole pair leads to the Elliot-formula containing the exciton peak, as well as the Coulomb-enhancement. Some discussion is devoted also to a modern branch of optical experiments with ultra-short and intense laser beams. The so-called semiconductor Bloch equations are derived and a third order non-linear approach to differential transmission (DTS) and four wave mixing (FWM) is developed.

6.1 Linear Response to a Time-Dependent External Perturbation

In this chapter we shall discuss the interaction of quantum mechanical charged particle systems with external electromagnetic fields in the optical range (light). The treatment first will be restricted to phenomena in the linear domain and we shall develop the conventional formalism for this purpose. Later we shall treat also some topics of non-linear optics related to the interaction of matter with intense laser pulses.

L. A. Bányai, *A Compendium of Solid State Theory*,
https://doi.org/10.1007/978-3-030-37359-7_6

Let us consider a quantum mechanical system defined by the Hamilton operator H. We want to look for the response of this system to a weak time-dependent external perturbation $H'(t)$. The total Hamiltonian we consider is

$$H_{tot}(t) = H + H'(t).$$

The state of a quantum mechanical statistical ensemble is described by the density operator $R(t)$ and the average $\langle A \rangle$ of an observable A is given by $Tr\{R(t)A\}$. The density operator is Hermitian, positive definite $R(t) \geq 0$ (its eigenvalues are non-negative) and is normalized $Tr\{R(t)\} = 1$. The time evolution is governed by the Liouville equation

$$\imath\hbar \frac{\partial R(t)}{\partial t} = [H_{tot}(t), R(t)].$$

In the interaction picture defined by the transformation

$$\bar{R}(t) = e^{\frac{\imath H(t-t_0)}{\hbar}} R(t) e^{-\frac{\imath H(t-t_0)}{\hbar}}$$

the Liouville equation looks as

$$\imath\hbar \frac{\partial \bar{R}}{\partial t} = \left[\bar{H}'(t), \bar{R}\right]; \qquad \bar{H}'(t) = e^{\frac{\imath H(t-t_0)}{\hbar}} H'(t) e^{-\frac{\imath H(t-t_0)}{\hbar}}.$$

Let us consider that at some time t_0 before the introduction of the perturbation the system was in thermal equilibrium

$$R(t_0) = R_0 \equiv \frac{e^{-\beta(H-\mu N)}}{Tr\left\{e^{-\beta(H-\mu N)}\right\}}.$$

Then for the deviation from the equilibrium

$$\bar{R} = R_0 + \delta\bar{R}$$

we have

$$\imath\hbar \frac{\partial \delta\bar{R}(t)}{\partial t} = \left[\bar{H}'(t), R_0\right] + \dots.$$

where the unspecified terms must be of higher order in the perturbation. The formal solution with the initial condition

$$\delta\bar{R}(t_0) = 0$$

is

$$\delta\bar{\rho}(t) = \frac{1}{\imath\hbar}\int_{t_0}^{t} dt' \left[\bar{H}'(t'), \rho_0\right] + \ldots$$

or

$$R(t) = R_0 + \frac{1}{\imath\hbar}\int_{t_0}^{t} dt' e^{\frac{\imath H(t'-t)}{\hbar}} \left[H'(t'), R_0\right] e^{\frac{-\imath H(t'-t)}{\hbar}} + \ldots$$

We are interested here only in terms linear in the perturbation and consequently, omit all higher order terms. Then the average of an observable A is given by

$$\begin{aligned}\langle A(t)\rangle - \langle A\rangle_0 &= \frac{1}{\imath\hbar}\int_{t_0}^{t} dt' Tr\left\{e^{\frac{\imath H(t'-t)}{\hbar}} \left[H'(t'), R_0\right] e^{\frac{-\imath H(t'-t)}{\hbar}} A\right\} \\ &= \frac{1}{\imath\hbar}\int_{t_0}^{t} dt' \langle\left[A_H(t-t'), H'(t')\right]\rangle_0,\end{aligned}$$

where we used cyclical permutability under the trace and $\langle\ldots\rangle_0$ denotes the equilibrium average. This defines the linear response of the system to a time-dependent perturbation.

6.2 Equilibrium Linear Response

There is another kind of linear response problem that we describe now. Let us suppose that the system is in equilibrium in the presence of a time-independent external perturbation. An example of such a situation we considered earlier was the penetration of a static electric field into a semiconductor. Here we treat the general case in order to underline the difference to the previous time-dependent non-equilibrium problem.

Thus, we consider a system described by a time-independent Hamiltonian, containing a small perturbation H'

$$H_{tot} = H + H'$$

in a state described by the macro-canonical equilibrium density operator

$$R = \frac{e^{-\beta(H_{tot}-\mu N)}}{Tr\left\{e^{-\beta(H_{tot}-\mu N)}\right\}}.$$

We want to find out the deviation of the equilibrium average of a certain observable A in the presence of the perturbation from its equilibrium average in the absence of the perturbation

$$\delta\langle A\rangle = \langle A\rangle - \langle A\rangle_0 = Tr\{RA\} - Tr\{R_0 A\}\ ,$$

where

$$R = \frac{e^{-\beta(H_{tot}-\mu N)}}{Tr\left\{e^{-\beta(H_{tot}-\mu N)}\right\}};\quad R_0 = \frac{e^{-\beta(H-\mu N)}}{Tr\left\{e^{-\beta(H-\mu N)}\right\}}.$$

We develop the exponential operator in powers of the perturbation H' using

$$e^{-\beta(H+H'-\mu N)} = e^{-\beta(H-\mu N)}\left\{1 - \int_0^\beta d\lambda e^{\lambda(H-\mu N)} H' e^{-\lambda(H-\mu N)} + \ldots\right\}$$

to get the equilibrium linear response

$$\delta\langle A\rangle = \int_0^\beta d\lambda\langle\left[H'(-\imath\hbar\lambda) - \langle H'\rangle_0\right](A - \langle A\rangle_0)\rangle_0. \tag{6.1}$$

There are two important remarks to be made here:

(i) We admitted implicitly that the chemical potential μ and the inverse temperature β remain unchanged by the introduction of the perturbation; however, this implies that the average energy and the average number of particles is not the same. Alternatively, we could have fixed the average particle and energy densities and include the corresponding variation of the chemical potential and temperature.
(ii) It is not at all obvious that the zero-frequency limit of the adiabatic time-dependent linear response coincides with the equilibrium linear response result.

The origin of this discrepancy lies in the delicate problem of irreversibility. Only irreversible processes can bring a system to equilibrium and these are not included in the quantum mechanical evolution. See the previous discussion of irreversibility in Chap. 5.

6.3 Dielectric Response of a Coulomb Interacting Electron System

Let us now consider the perturbation of a Coulomb interacting electron system (with a uniform compensating positive background) by an external longitudinal electric field defined by an external potential $V^{ext}(\mathbf{x}, t)$. Interactions with phonons may be included. In equilibrium the average charge density $\langle\rho(\mathbf{x})\rangle_0$ is supposed to be vanishing everywhere, while due to the perturbation

$$H'(t) = \int d\mathbf{x}\rho(\mathbf{x})V^{ext}(\mathbf{x}, t)$$

it will be different from zero. By pushing $t_0 \to -\infty$, the linear response to this perturbation is

$$\langle \rho(\mathbf{x}, t) \rangle = \int_{-\infty}^{\infty} dt' \int d\mathbf{x}' \kappa(\mathbf{x}, \mathbf{x}'; t - t') V^{ext}(\mathbf{x}', t')$$

with the kernel being the charge density correlation function

$$\kappa(\mathbf{x}, \mathbf{x}'; t) \equiv \frac{\theta(t)}{\imath\hbar} \langle [\rho(\mathbf{x}, t)_H , \rho(\mathbf{x}', 0)_H] \rangle_0 .$$

In our homogeneous system, this function may depend only on the difference of the coordinates $\kappa(\mathbf{x}, \mathbf{x}'; t) \equiv \kappa(\mathbf{x} - \mathbf{x}'; t)$. While extending the integration in time to $-\infty$, one has to consider the introduction of the perturbation in an adiabatic manner

$$V^{ext}(\mathbf{x}, t) = e^{\imath(\omega + \imath 0)t} V^{ext}(\mathbf{x}) ,$$

in order to have well defined Fourier transforms. The symbol $+\imath 0$ here means that one considers a small positive adiabatic parameter that after the performed integration goes to zero.
After a Fourier transformation in the time and space variables we get

$$\langle \tilde{\rho}(\mathbf{k}, \omega) \rangle = \tilde{\kappa}(\mathbf{k}, \omega) \tilde{V}^{ext}(\mathbf{k}, \omega)$$

with

$$\tilde{\kappa}(\mathbf{k}, \omega) = \int_{0}^{\infty} dt \int d\mathbf{x} e^{-\imath \mathbf{k}\mathbf{x}} e^{\imath(\omega + \imath 0)t} \kappa(\mathbf{x}, t).$$

Now, it is very important to realize that in a Coulomb system the local charge density at its turn is the source of an internal potential. We are interested in the relation between the internal charge density and the total potential $V = V^{ext} + V^{int}$ that is defined by the Poisson equation (here in Fourier transforms)

$$k^2 \tilde{V}(\mathbf{k}, \omega) = 4\pi \langle \tilde{\rho}(\mathbf{k}, \omega) \rangle + 4\pi \rho^{ext}(\mathbf{k}, \omega)$$

while

$$k^2 \tilde{V}^{ext}(\mathbf{k}, \omega) = 4\pi \rho^{ext}(\mathbf{k}, \omega).$$

From these equations and the linear response of the charge density to the external potential we get

$$\epsilon(\mathbf{k}, \omega) k^2 \tilde{V}(\mathbf{k}, \omega) = 4\pi \rho^{ext}(\mathbf{k}, \omega)$$

with the dielectric function

$$\epsilon(\mathbf{k},\ \omega) \equiv \frac{1}{1 + \frac{4\pi}{k^2}\tilde{\kappa}(\mathbf{k}, \omega)}. \tag{6.2}$$

Obviously due to the relationship between the conductivity and the dielectric function

$$\sigma(\mathbf{k}, \omega) = -\imath\omega\epsilon(\mathbf{k}, \omega) \tag{6.3}$$

to get a finite d.c. conductivity $\sigma_{dc} \equiv \sigma(0, 0)$ one needs a diverging imaginary part of the dielectric function by $\omega \to 0$. The situation is quite different if one treats the Coulomb interacting electron plasma in the s.c. Hartree approximation. Then the perturbation is the self-consistent potential itself and not the external one. Therefore, from the Poisson equation one gets

$$\epsilon(\mathbf{k},\ \omega)^{sc} \equiv 1 - \frac{4\pi}{k^2}\tilde{\kappa}^{sc}(\mathbf{k}, \omega). \tag{6.4}$$

However, the two density correlation functions would be quite different!

6.4 The Full Nyquist Theorem

To stress the importance of the conclusions of the previous Sect. 6.3 concerning the interpretation of the linear dielectric response in the presence of Coulomb interactions we describe here the derivation of the Nyquist theorem. This theorem was deduced within the classical thermodynamics of electric circuitry. It is a statement relating the thermal fluctuation (noise) of the potential drop U in a conductor to its resistance R and temperature T:

$$\langle \delta U^2 \rangle_0 = R k_B T.$$

Since the potential drop along the conductor is given by the field strength E along the wire multiplied by the length L of the wire $U = EL$ and the resistance is related to the d.c. conductance σ_{dc}, the length L and cross section S of the wire by $R = \frac{L}{\sigma_{dc} S}$, one gets

$$\langle \delta E^2 \rangle_0 = \frac{k_B T}{\Omega \sigma_{dc}}, \tag{6.5}$$

where $\Omega = LS$ is the volume of the macroscopic wire. This later form is suitable for a microscopical treatment.

The spectral density of the equilibrium quantum mechanical noise of the (Hermitian) observable A (whose equilibrium average is vanishing) is defined as:

$$\delta A_{\omega}^{2} \equiv \frac{1}{2}\int_{0}^{\infty} dt\cos(\omega t)\langle\{AA(t)+A(t)A\}\rangle_{0}, \tag{6.6}$$

with the notations

$$A(t)=e^{\frac{\imath}{\hbar}Ht}Ae^{-\frac{\imath}{\hbar}Ht};\ \langle\ldots\rangle_{0}\equiv Tr\left(\frac{1}{Z}e^{-\beta(H-\mu N)}\ldots\right).$$

Here we have assumed that the correlations vanish as $t\rightarrow\infty$ that implies the inclusion of the interaction with some thermal bath (for example phonons).

After some transformations Eq. 6.6 may be brought to the form

$$\delta A_{\omega}^{2}=\frac{1}{2}\coth\left(\frac{\beta\hbar\omega}{2}\right)\Re\left\{\int_{0}^{\infty} dte^{-\imath\omega t}\langle[A,A(t)]\rangle_{0}\right\}. \tag{6.7}$$

Now, the microscopical field (operator) is

$$\mathbf{E}(\mathbf{x},t)=-\nabla V(\mathbf{x},t),$$

with

$$V(\mathbf{x},t)=V_{ext}(\mathbf{x},t)+\int d\mathbf{x}'\frac{\rho(\mathbf{x}',t)}{|\mathbf{x}-\mathbf{x}'|},$$

expressed through the operator of the particle density $\rho(\mathbf{x},t)$. Since the external potential V_{ext} is a c-number, the fluctuation of the field is completely determined by the fluctuation of the internal field

$$\mathbf{E}_{int}(\mathbf{x},t)=-\nabla\int d\mathbf{x}'\frac{\rho(\mathbf{x}',t)}{|\mathbf{x}-\mathbf{x}'|}.$$

After a Fourier transform in space

$$\tilde{V}(\mathbf{k},t)\equiv\int d\mathbf{x}e^{-\imath\mathbf{kx}}V(\mathbf{x},t),$$

one has

$$\tilde{\mathbf{E}}_{int}(\mathbf{k},t)=\imath\mathbf{k}V(\mathbf{k},t)=\imath\mathbf{k}\frac{4\pi}{k^{2}}\rho(\mathbf{k},t).$$

Let us now define

$$\langle\langle \tilde{\mathbf{E}}_{int}(\mathbf{k})\tilde{\mathbf{E}}_{int}(-\mathbf{k})\rangle\rangle_\omega \equiv \frac{1}{2}\coth\left(\frac{\beta\hbar\omega}{2}\right)\Re\left\{\int_0^\infty dt e^{-\imath\omega t}\langle\left[\tilde{\mathbf{E}}_{int}(\mathbf{k}), \tilde{\mathbf{E}}_{int}(-\mathbf{k}, t)\right]\rangle_0\right\},$$

(scalar product of vectors is understood), or

$$\langle\langle \mathbf{E}_{int}(\mathbf{k})\mathbf{E}_{int}(-\mathbf{k})\rangle\rangle_\omega = \frac{1}{2}\coth\left(\frac{\beta\hbar\omega}{2}\right)\frac{(4\pi)^2}{k^2}\Re\left\{\int_0^\infty dt e^{-\imath\omega t}\langle\left[\tilde{\rho}(\mathbf{k}), \tilde{\rho}(-\mathbf{k}, t)\right]\rangle_0\right\}.$$

On the other hand due to translational invariance

$$\langle\left[\tilde{\rho}(\mathbf{k}), \tilde{\rho}(-\mathbf{k}, t)\right]\rangle_0 = \Omega \int d\mathbf{x} e^{-\imath\mathbf{kx}}\langle[\rho(\mathbf{x}), \delta\rho(0, t)]\rangle_0$$

holds. Therefore,

$$\langle\langle \mathbf{E}_{int}(\mathbf{k})\mathbf{E}_{int}(-\mathbf{k})\rangle\rangle_\omega = \coth\left(\frac{\beta\hbar\omega}{2}\right)\frac{(4\pi)^2}{2k^2}\Omega \times \Re\left\{\int_0^\infty dt e^{-\imath\omega t}\int d\mathbf{x} e^{-\imath\mathbf{kx}}\langle[\rho(\mathbf{x}), \delta\rho(0, t)]\rangle_0\right\},$$

or

$$\langle\langle \mathbf{E}_{int}(\mathbf{k})\mathbf{E}_{int}(-\mathbf{k})\rangle\rangle_\omega = \coth\left(\frac{\beta\hbar\omega}{2}\right)\frac{(4\pi)^2}{2k^2}\Omega\Re\left\{\imath\hbar\tilde{\kappa}(\mathbf{k}, \omega)\right\},$$

results.

Using the relationship Eq. 6.2

$$\frac{4\pi}{k^2}\tilde{\kappa}(\mathbf{k}, \omega) = \frac{1}{\epsilon(\mathbf{k}, \omega)} - 1,$$

valid in the presence of **Coulomb interactions** in the Hamiltonian, we get

$$\langle\langle \mathbf{E}_{int}(\mathbf{k})\mathbf{E}_{int}(-\mathbf{k})\rangle\rangle_\omega = -\frac{1}{2}\coth\left(\frac{\beta\hbar\omega}{2}\right)4\pi\hbar\Omega\Im\left\{\frac{1}{\epsilon(\mathbf{k}, \omega)} - 1\right\}.$$

Since $\tilde{\mathbf{E}}_{int}(0) = \tilde{\mathbf{E}}_{int}(0)^+$, it follows that for $\mathbf{k} \to 0$ we have

$$\langle\langle \mathbf{E}_{int}(0)^2\rangle\rangle_\omega = -\frac{1}{2}\coth\left(\frac{\beta\hbar\omega}{2}\right)4\pi\hbar\Omega\Im\left\{\frac{1}{\epsilon(0, \omega)}\right\}. \quad (6.8)$$

On the other hand,

$$\Im\left\{\frac{1}{\epsilon(0, \omega)}\right\} = \frac{-\Im\epsilon(0, \omega)}{\Re\epsilon(0, \omega)^2 + \Im\epsilon(0, \omega)^2},$$

and since (see Eq. 6.3) at $\omega \to 0$ only the imaginary part of the dielectric function diverges as $\frac{4\pi}{\omega}\sigma_{dc}$, we get

$$\lim_{\omega \to 0} \langle\langle \mathbf{E}_{int}(0)^2 \rangle\rangle_\omega = \frac{k_B T \Omega}{\sigma_{dc}}. \tag{6.9}$$

If one interprets the classical analog of this quantum mechanical fluctuation in the previous equation as $\Omega^2 \langle \delta E^2 \rangle_0$, then we get the Nyquist theorem Eq. 6.5.

Actually from Eq. 6.8 emerges the full, frequency-dependent Nyquist theorem for the equilibrium field fluctuation

$$\langle \delta E_\omega^2 \rangle_0 = \frac{\omega}{2\Omega} \coth\left(\frac{\beta \hbar \omega}{2}\right) 4\pi \hbar \Re \left\{ \frac{1}{\sigma(0,\omega)} \right\}. \tag{6.10}$$

As we have seen, the essential point in the derivation of the Nyquist theorem was the correct treatment of the electromagnetic potential, taking into account that it is a dynamical variable determined by the internal charge and the use of the correct definition of the dielectric function Eq. 6.2 in the presence of Coulomb interactions.

6.5 Dielectric Function of an Electron Plasma in the Hartree Approximation

As we have seen before, in the self-consistent Hartree approximation one has free electrons with modified s.c. one-particle energies. In what follows we ignore this modification and therefore our electrons in the absence of the electric field are treated just to be free. The charge density operator in terms of the second quantized wave functions is

$$\rho(\mathbf{x}) = e\psi(\mathbf{x})^+ \psi(\mathbf{x}).$$

Here we ignored the positive uniform compensating charge that plays no role in the following discussion. In terms of the creation–annihilation operators of the one-particle states of wave vector **k**, in the Heisenberg picture

$$\rho(\mathbf{r},t)_H = \frac{e}{\Omega} \sum_{\mathbf{k}} \sum_{\mathbf{k}'} c_{\mathbf{k}}^+ c_{\mathbf{k}'} e^{\imath(\mathbf{k}'-\mathbf{k})\mathbf{r}} e^{\frac{\imath}{\hbar}(e_{\mathbf{k}}-e_{\mathbf{k}'})t}$$

with the kinetic energies $e_{\mathbf{k}} = \frac{\hbar^2 k^2}{2m}$. The density correlation function is

$$\kappa^{sc}(\mathbf{r},t) = \frac{e^2}{\imath\hbar\Omega^2} \sum_{\mathbf{k}} \sum_{\mathbf{k}'} \sum_{\mathbf{p}} \sum_{\mathbf{p}'} e^{\frac{\imath}{\hbar}(e_{\mathbf{k}}-e_{\mathbf{k}'})t} e^{\imath(\mathbf{k}'-\mathbf{k})\mathbf{r}} \langle \left[c_{\mathbf{k}}^+ c_{\mathbf{k}'}, c_{\mathbf{p}}^+ c_{\mathbf{p}'} \right] \rangle_0.$$

The average $\langle\left[c_{\mathbf{k}}^{+}\, c_{\mathbf{k}'}\,,\, c_{\mathbf{p}}^{+}\, c_{\mathbf{p}'}\right]\rangle_0$ may be easily performed by knowing the commutator

$$\left[c_{\mathbf{k}}^{+}\, c_{\mathbf{k}'}\,,\, c_{\mathbf{p}}^{+}\, c_{\mathbf{p}'}\right] = \delta_{\mathbf{k}'\,\mathbf{p}}\, c_{\mathbf{k}}^{+}\, c_{\mathbf{p}'} - \delta_{\mathbf{p}'\,\mathbf{k}}\, c_{\mathbf{p}}^{+}\, c_{\mathbf{k}'}$$

and retaining the only surviving average

$$\langle c_{\mathbf{k}}^{+}\, c_{\mathbf{p}'}\rangle_0 = \delta_{\mathbf{k}\,\mathbf{p}'}\, f_{\mathbf{k}}\,,$$

with $f_{\mathbf{k}} \equiv \left\langle c_{\mathbf{k}}^{+}\, c_{\mathbf{k}}\right\rangle_0$ being the Fermi function. One gets finally

$$\epsilon_L(\mathbf{q}, \omega) = 1 - \frac{4\pi e^2}{\Omega q^2} \sum_{\mathbf{k}} \frac{f_{\mathbf{k}+\mathbf{q}} - f_{\mathbf{k}}}{e_{\mathbf{k}+\mathbf{q}} - e_{\mathbf{k}} - \hbar\omega - \imath 0}.$$

In the evaluation of this formula one has to take into account the identity

$$\frac{1}{x - \imath 0} = P\left(\frac{1}{x}\right) + \imath\pi\,\delta(x)$$

with the first term being the principal value, while the second one contains Dirac's delta function. Then

$$\Re\epsilon_L(\mathbf{q}, \omega) = 1 - \frac{4\pi e^2}{\Omega q^2} \sum_{\mathbf{k}} \frac{2 f_{\mathbf{k}}(e_{\mathbf{k}} - e_{\mathbf{k}+\mathbf{q}})}{(e_{\mathbf{k}} - e_{\mathbf{k}+\mathbf{q}})^2 + (\hbar\omega)^2}$$

and

$$\begin{aligned}\Im\epsilon_L(\mathbf{q}, \omega) &= \frac{4\pi^2 e^2}{\Omega q^2} \sum_{\mathbf{k}} \left(f_{\mathbf{k}} - f_{\mathbf{k}+\mathbf{q}}\right) \delta(\hbar\omega + e_{\mathbf{k}} - e_{\mathbf{k}+\mathbf{q}}) \\ &= \frac{4\pi^2 e^2}{\Omega q^2} \sum_{\mathbf{k}} \left(f(e_{\mathbf{k}}) - f(e_{\mathbf{k}} + \hbar\omega)\right) \delta(\hbar\omega + e_{\mathbf{k}} - e_{\mathbf{k}+\mathbf{q}}) \\ &= \frac{e^2 m^2}{\hbar^4 q^3} \int_0^\infty de\, (f(e) - f(e + \hbar\omega))\, \theta\left(1 - \frac{\left|\hbar\omega - \frac{\hbar^2 q^2}{2m}\right|}{2\sqrt{e\frac{\hbar^2 q^2}{2m}}}\right).\end{aligned}$$

At $T = 0^0 K$ we get

$$\Im\epsilon_L(\mathbf{q}, \omega) = \frac{e^2 m^2}{\hbar^4 q^3} \int_{E_F - \hbar\omega}^{E_F} de\theta\left(1 - \frac{\left|\hbar\omega - \frac{\hbar^2 q^2}{2m}\right|}{2\sqrt{e\frac{\hbar^2 q^2}{2m}}}\right)$$

and the imaginary part of the dielectric function vanishes for

$$\left(\hbar\omega - \frac{\hbar^2 q^2}{2m}\right)^2 > \frac{2\hbar^2 q^2}{m} E_F.$$

At any temperature asymptotically for $\omega \to \infty$

$$\Im\epsilon_L(\mathbf{q}, \omega) \approx 0$$

and

$$\Re\epsilon_L(\mathbf{q}, \omega) \approx 1 - \frac{4\pi e^2 \langle n \rangle}{m\omega^2} = 1 - \left(\frac{\omega_{pl}}{\omega}\right)^2$$

and

$$\omega_{pl} = \sqrt{\frac{4\pi e^2 \langle n \rangle}{m}}$$

is the plasma frequency. Since the dielectric constant vanishes for big $\omega = \omega_{pl}$ (provided ω_{pl} is also big!), eigenoscillations of the plasma may occur in the absence of an external field at this frequency.
Peculiar properties has also the static longitudinal dielectric function

$$\epsilon_L(\mathbf{q}, 0) = 1 - \frac{4\pi e^2}{\Omega q^2} \sum_{\mathbf{k}} P\left(\frac{f_{\mathbf{k}+\mathbf{q}} - f_{\mathbf{k}}}{e_{\mathbf{k}+\mathbf{q}} - e_{\mathbf{k}}}\right).$$

For $\mathbf{q} \to 0$

$$\epsilon_L(\mathbf{q}, 0) \approx 1 - \frac{4\pi e^2}{\Omega q^2} \sum_{\mathbf{k}} \frac{\partial f(e_{\mathbf{k}})}{\partial e_{\mathbf{k}}} = 1 - \frac{4\pi e^2}{q^2} \int dez(e) \frac{\partial f(e)}{\partial e} = 1 + \frac{\kappa^2}{q^2}$$

with

$$\kappa^2 \equiv 4\pi e^2 \int dez(e) \frac{\partial f(e)}{\partial e} \approx \begin{cases} \frac{4\pi}{k_B T} \langle n \rangle \ (T \to \infty) \\ \frac{6\pi e^2}{E_F} \langle n \rangle \ (T \to 0) \end{cases}$$

This last result shows that if one puts an external point-like charge q inside the plasma ($V^{ext}(\mathbf{r}) = \frac{q}{r}$ i.e. $\tilde{V}^{ext}(\mathbf{q}) = \frac{4\pi}{q^2}$), then the resulting potential $\tilde{V}(q)$ for small q looks as

$$\tilde{V}(\mathbf{q}) = \frac{\tilde{V}^{ext}(\mathbf{q})}{\epsilon_L(\mathbf{q}, 0)} \approx \frac{4\pi e^2}{q^2 + \kappa^2}$$

and at big distances in the real space it will look as a Yukawa potential

$$V(\mathbf{r}) \approx \frac{e^2}{|\mathbf{r}|} e^{-\kappa r}.$$

It is important to remark that in the whole discussion here no interaction with any thermal bath was considered and therefore there were no irreversible effects incorporated.

6.6 The Transverse, Inter-Band Dielectric Response of an Electron–Hole Plasma

A transverse propagating electric field $\mathbf{E}(\mathbf{x}, t)$ (photon) may be defined by the choice of a vanishing scalar potential ($V(\mathbf{x}, t) = 0$) and a transverse vector potential ($\nabla \mathbf{A}(\mathbf{x}, t) = 0$). The interaction of charged particles with this electromagnetic field (ignoring the $\mathbf{A}^2$ term) is given by

$$-\int d\mathbf{x} \mathbf{j}(\mathbf{x}) \mathbf{A}(\mathbf{x}, t).$$

In a direct gap semiconductor, due to its vector character, this interaction may couple mainly s-like states of the conduction band with p-like states of the degenerate valence band.

In an electron–hole plasma model, therefore, one describes such an electromagnetic interaction with a dipole Hamiltonian

$$H_{em}(t) = d \sum_{\mu=1}^{3} \sum_{\sigma=\pm\frac{1}{2}} \int d\mathbf{x} \psi_{e,-\sigma}(\mathbf{x})^{+} \psi_{h,\mu,\sigma}(\mathbf{x})^{+} \mathscr{E}_{\mu}(\mathbf{x}, t) + h.c..$$

The second quantized wave functions of the holes $\psi_{h,\mu,\sigma}(\mathbf{x})$ correspond to the three p-type valence bands of spin σ. Here d is a phenomenological dipole constant and the spin is $\sigma = \pm\frac{1}{2}$. The above Hamiltonian containing pairs of creation and annihilation operators indicates that a photon having an energy $\hbar\omega$ above the band gap may create an electron–hole pair.

We want to take into account here that the newly created electron and hole interact through Coulomb forces. As we shall see, this fact has important consequences for the absorption spectrum. The Coulomb interaction we shall treat within the intra-band s.c. Fock approximation by retaining the non-local average (at instant t)

$$\langle \psi_{e,\sigma}(\mathbf{x}) \psi_{h,\mu,-\sigma}(\mathbf{x}') \rangle_t .$$

Such an average vanishes in the absence of the electric field, but appears after the introduction of the field.
Within this approximation one has an s.c. Coulomb energy

$$\sum_{\mu=1}^{3}\sum_{\sigma=\pm\frac{1}{2}}\int d\mathbf{x}\int d\mathbf{x}'\frac{e^2}{|\mathbf{x}-\mathbf{x}'|}\left(\psi_{e,\sigma}(\mathbf{x})\psi_{h,\mu,-\sigma}(\mathbf{x}')\langle\psi_{e,\sigma}(\mathbf{x})\psi_{h,\mu,-\sigma}(\mathbf{x}')\rangle_t^* + h.c.\right).$$

Obviously, this term has the same operator structure as $H_{em}(t)$ and is at least of first order in the field. Therefore, the effective perturbation consists of $H_{em}(t)$ plus an induced perturbation of the same structure

$$H'(t) = d\sum_{\mu=1}^{3}\int d\mathbf{x}\mathscr{P}_\mu(\mathbf{x},\mathbf{x})\mathscr{E}_\mu(\mathbf{x},t)$$
$$+\sum_{\mu=1}^{3}\int d\mathbf{x}\int d\mathbf{x}'\frac{e^2}{|\mathbf{x}-\mathbf{x}'|}\mathscr{P}_\mu(\mathbf{x},\mathbf{x}')\langle\mathscr{P}_\mu(\mathbf{x},\mathbf{x}')\rangle_t^* + h.c.,$$

where the notation

$$\mathscr{P}_\mu(\mathbf{x},\mathbf{x}') \equiv \sum_{\sigma=\pm\frac{1}{2}}\psi_{e,\sigma}(\mathbf{x})\psi_{h,\mu,-\sigma}(\mathbf{x}')$$

was introduced.
Let us consider now the Heisenberg equations of motion for the operator $\mathscr{P}_\mu(\mathbf{x},\mathbf{x}')$. Using the commutation relation

$$\left[\psi_{e,\sigma}(\mathbf{x})\psi_{h_\mu,-\sigma}(\mathbf{x}'),\psi_{h_\nu,-\sigma}(\mathbf{y}')^+\psi_{e,\sigma}(\mathbf{y})^+\right] = \delta_{\mu\nu}\delta(\mathbf{x}-\mathbf{y})\delta(\mathbf{x}'-\mathbf{y}')$$
$$-\delta(\mathbf{x}-\mathbf{y})\psi_{h,-\sigma}(\mathbf{y}')^+\psi_{h,-\sigma}(\mathbf{x}') - \delta(\mathbf{x}'-\mathbf{y}')\delta_{\mu\nu}\psi_{e,\sigma}(\mathbf{y})^+\psi_{e,\sigma}(\mathbf{x})$$

we get

$$\left\{\imath\hbar\frac{\partial}{\partial t} + \frac{\hbar^2}{2m_e}\nabla_e^2 + \frac{\hbar^2}{2m_h}\nabla_h^2 + \frac{e^2}{|\mathbf{x}_e-\mathbf{x}_h|}\right\}\mathscr{P}_\mu(\mathbf{x}_e,\mathbf{x}_h) =$$
$$d\sum_{\nu=1}^{3}\sum_{\sigma}\int d\mathbf{y}\mathscr{E}_\nu(\mathbf{y},t)\left(\delta_{\mu\nu}\delta(\mathbf{x}_e-\mathbf{y})\delta(\mathbf{x}_h-\mathbf{y})\right.$$
$$\left.-\delta_{\mu\nu}\delta(\mathbf{x}_h-\mathbf{y})\psi_{e,\sigma}(\mathbf{x}_h)^+\psi_{e,\sigma}(\mathbf{x}_e) - \delta(\mathbf{x}_e-\mathbf{y})\psi_{h_\mu,-\sigma}(\mathbf{x}_e)^+\psi_{h_\nu,-\sigma}(\mathbf{x}_h)\right)$$
$$-\sum_{\nu=1}^{3}\sum_{\sigma}\int d\mathbf{x}\int d\mathbf{x}'\frac{e^2}{|\mathbf{x}-\mathbf{x}'|}\langle\mathscr{P}_\nu(\mathbf{x},\mathbf{x}')\rangle_t\left(\delta_{\mu\nu}\delta(\mathbf{x}-\mathbf{x}_e)\delta(\mathbf{x}'-\mathbf{x}_h)\right.=$$

$$-\delta_{\mu\nu}\delta(\mathbf{x}_h - \mathbf{x}')\psi_{e,\sigma}(\mathbf{x})^+\psi_{e,\sigma}(\mathbf{x}_e) - \delta(\mathbf{x}_e - \mathbf{x})\psi_{h_\mu,-\sigma}(\mathbf{x}')^+\psi_{h_\nu,-\sigma}(\mathbf{x}_h)\big) .$$

Now, we perform the average $\langle \ldots \rangle_t$ by taking into account that in the absence of the field there were no electrons and holes. The initial state is the electron–hole vacuum and even thereafter the population appears only in second order in the field. Therefore we may consider

$$\langle \psi_{\alpha,\sigma}(\mathbf{x})^+\psi_{\alpha,\sigma}(\mathbf{x}')\rangle_t \approx 0; \quad (\alpha = e, h).$$

In the linear approximation in the external field we get therefore the equation of motion

$$\left\{\imath\hbar\frac{\partial}{\partial t}+\frac{\hbar^2}{2m_e}\nabla_e^2+\frac{\hbar^2}{2m_h}\nabla_h^2+\frac{e^2}{|\mathbf{x}_e - \mathbf{x}_h|}+E_g\right\}\langle \mathscr{P}_\mu(\mathbf{x}_e, \mathbf{x}_h)\rangle_t = d\mathscr{E}_\mu(\mathbf{x}_e, t)\delta(\mathbf{x}_e - \mathbf{x}_h),$$

where E_g is the band gap energy.

The solution of this non-homogeneous equation one gets by using the Green function

$$G(\mathbf{x}_e, \mathbf{x}_h; \mathbf{x}'_e, \mathbf{x}'_h; t)$$

of the electron–hole Schrödinger equation

$$\left\{\imath\hbar\frac{\partial}{\partial t}+\frac{\hbar^2}{2m_e}\nabla_e^2+\frac{\hbar^2}{2m_h}\nabla_h^2+\frac{e^2}{|\mathbf{x}_e - \mathbf{x}_h|}\right\}G(\mathbf{x}_e, \mathbf{x}_h; \mathbf{x}'_e, \mathbf{x}'_h; t) =$$

$$\delta(\mathbf{x}_e - \mathbf{x}'_e)\delta(\mathbf{x}_h - \mathbf{x}'_h)\delta(t).$$

On its turn the Green function may be built up from the eigenfunctions $\Phi_\alpha(\mathbf{x}_e, \mathbf{x}_h)$ of the Coulomb interacting electron–hole Schrödinger equation

$$\left\{-\frac{\hbar^2}{2m_e}\nabla_e^2 - \frac{\hbar^2}{2m_h}\nabla_h^2 - \frac{e^2}{|\mathbf{x}_e - \mathbf{x}_h|}\right\}\Phi_\alpha(\mathbf{x}_e, \mathbf{x}_h) = \varepsilon_\alpha\Phi_\alpha(\mathbf{x}_e, \mathbf{x}_h)$$

as

$$G_r(\mathbf{x}_e, \mathbf{x}_h; \mathbf{x}'_e, \mathbf{x}'_h; t) = \frac{1}{\imath\hbar}\theta(t)\sum_\alpha \Phi_\alpha(\mathbf{x}_e, \mathbf{x}_h)_\alpha\Phi_\alpha(\mathbf{x}'_e, \mathbf{x}'_h)^* e^{\imath\frac{\varepsilon_\alpha}{\hbar}t}.$$

Here we considered the retarded Green function relevant for our physical problem. One may separate the center of mass motion

$$\mathbf{X} \equiv \frac{m_e\mathbf{x}_e + m_h\mathbf{x}_h}{m_e + m}$$

from the relative motion and

$$\Phi_{\mathbf{q},l,m,\varepsilon}(\mathbf{x}_e,\mathbf{x}_h) = \frac{1}{\sqrt{\Omega}} e^{-\imath \mathbf{q}\mathbf{X}} \phi_{l,m,\varepsilon}(\mathbf{x}) \,,$$

where $\phi_{l,m,\varepsilon}(\mathbf{x})$ are the well-known Coulomb wave functions characterized by their angular momentum l, m and energy ε while Ω is the volume of the system. These states may belong to the discrete spectrum or to the continuum.
Using the Green function $G(\mathbf{x}_e,\mathbf{x}_h;\mathbf{x}'_e,\mathbf{x}'_h;t)$ one finds

$$\langle \mathscr{P}_\mu(\mathbf{x}_e,\mathbf{x}_h)\rangle_t = \int_{-\infty}^{t} dt' \int d\mathbf{x}'_e \int d\mathbf{x}'_h e^{\frac{\imath}{\hbar}E_g(t-t')} G(\mathbf{x}_e,\mathbf{x}_h;\mathbf{x}'_e,\mathbf{x}'_h;t-t') d\mathscr{E}_\mu(\mathbf{x}'_e,t')\delta(\mathbf{x}'_e-\mathbf{x}'_h)$$

or

$$\langle \mathscr{P}_\mu(\mathbf{x},\mathbf{X})\rangle_t = \frac{d}{\imath\hbar\Omega} \sum_{\mathbf{q},l,m,\varepsilon} \int_{-\infty}^{t} dt' e^{\frac{\imath}{\hbar}(\varepsilon+E_g+\frac{\hbar^2 q^2}{2(m_e+m_h)})(t-t')}$$
$$\times \int d\mathbf{X}' e^{-\imath\mathbf{q}(\mathbf{X}-\mathbf{X}')} \phi_{l,m,\varepsilon}(\mathbf{x}) \phi_{l,m,\varepsilon}(0)^* \mathscr{E}_\mu(\mathbf{X}',t').$$

The transverse, inter-band dielectric function is defined by the relation of the inter-band polarization $\mathbf{P}(\mathbf{x},t) = d\langle\mathscr{P}(0,\mathbf{x})\rangle_t$ to the electric field

$$\mathbf{P}(\mathbf{q},\omega) = \frac{1}{4\pi}(\epsilon_T(\mathbf{q},\omega)-1)\mathbf{E}_t(\mathbf{q},\omega)$$

and we get

$$\epsilon_T(\mathbf{q},\omega) = 1 - 4\pi d^2 \sum_\varepsilon |\phi_{0,0,\varepsilon}(0)|^2$$
$$\times \left(\frac{1}{\hbar\omega - E_g - \varepsilon - \frac{\hbar^2 q^2}{2(m_e+m_h)} + \imath 0} - \frac{1}{\hbar\omega + E_g + \varepsilon + \frac{\hbar^2 q^2}{2(m_e+m_h)} + \imath 0} \right).$$

Here we took into account that in $\mathbf{x} = 0$ only the s-wave functions contribute.

The imaginary part that describes the absorption spectrum is given then explicitly by the Elliott formula

$$\Im\epsilon_T(\mathbf{q},\omega) = \frac{4d^2}{a_B^3 E_R}\left[\sum_{n=1}^{\infty} \frac{1}{n^3}\delta(\Delta+\frac{1}{n^2}) + \frac{1}{2}\theta(\Delta)\frac{1}{1-e^{-\frac{2\pi}{\sqrt{\Delta}}}}\right]$$

with

$$\Delta \equiv \frac{\hbar\omega - E_g - \frac{\hbar^2 q^2}{2(m_e+m_h)}}{E_R} ,$$

where E_R is the excitonic Rydberg energy.

In the absence of the Coulomb interaction ($e = 0$) we have had the simple interband absorption spectrum

$$\Im\epsilon_T(\mathbf{q}, \omega)^0 = \frac{d^2}{\pi a_B^3 E_R}\theta(\Delta)\sqrt{\Delta}.$$

In order to represent graphically in Fig. 6.1 the expected absorption spectrum we had to introduce by hand an element of irreversibility through an arbitrary line-width for the discrete exciton lines of which we retained only the two lowest states, but only the ground state may be actually seen. The Coulomb interaction between the electron and the hole created by photon absorption introduces the exciton resonance as well as a Coulomb-enhancement of the continuum at the threshold as to be compared with the red line of absorption without Coulomb effects. See also the comparison with the exciton state density Fig. 3.1 calculated within the tight binding Wannier scheme.

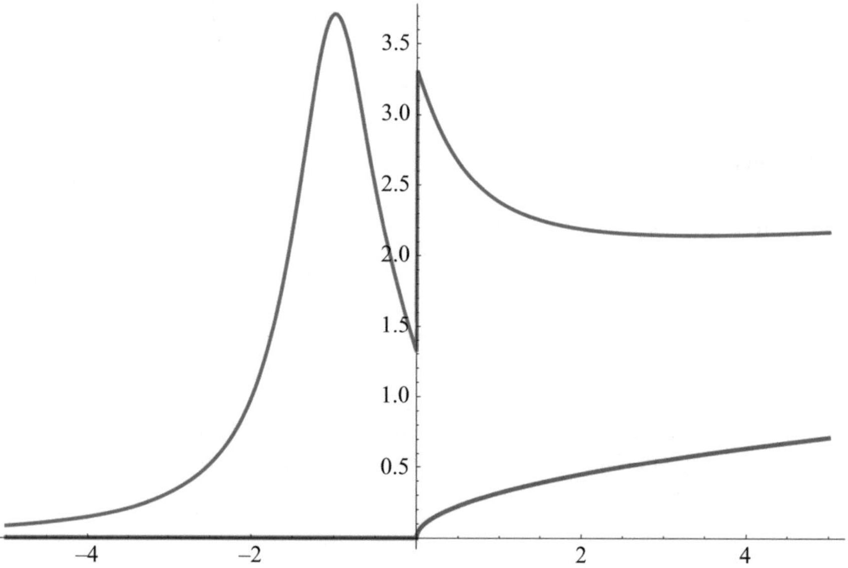

Fig. 6.1 Absorption spectra as function of the detuning Δ at the band edge ($\Delta = 0$) with Coulomb (blue) and without Coulomb effects (red)

6.7 Ultra-Short-Time Spectroscopy of Semiconductors

In the last decades interesting experiments were performed with very intense and ultra-short (on femtosecond scale!), laser beams on semiconductors. Their purpose is to obtain information about ultra-short time processes. These experiments are theoretically treated by the extension of the linear response including higher order terms in the development. However, drastic simplifications are needed. First of all, we shall ignore the coordinate dependence of the electric field **E** (negligible photon momentum!) and consider

$$\mathbf{E}(t) = \mathbf{E}(t)\cos(\omega_0 t)\,,$$

where ω_0 is the carrier frequency in the optical domain and $\mathbf{E}(t)$ is the pulse envelope. This may be justified by the fact that for photon energy in the neighborhood of the energy gap the photon wave length λ in the most interesting semiconductors is much bigger than the exciton radius a_B.

The system we consider is again an electron–hole plasma as in Sect. 6.6 with an s-like conduction band and a p-like valence band. The phenomenological coupling to the transverse electromagnetic field one uses is that of an electric dipole in analogy to the similar atomic coupling. We chose the quantization direction of the valence band states so that only one of the p bands couples to the field and, therefore, we may ignore the vector notation.

$$\begin{aligned} H_{em}(t) &= \hbar\omega_R(t)\cos(\omega_0 t)\sum_\sigma \int d\mathbf{x}\left(\psi_{e,\sigma}(\mathbf{x})^+\psi_{h,-\sigma}(\mathbf{x})^+ + h.c.\right) \\ &= \hbar\omega_R(t)\cos(\omega_0 t)\sum_{\sigma,\mathbf{k}}\left(a_{e,\sigma,\mathbf{k}}a_{h,-\sigma,-\mathbf{k}} + h.c.\right), \end{aligned}$$

where $\omega_R(t) = \frac{1}{\hbar}d\mathscr{E}(t)$ is the so-called (here, time dependent!) Rabi frequency. Obviously, higher order terms in the development of the system response would produce also oscillations with multiples of the frequency ω_0. Such terms may be less important if already $2\hbar\omega_0$ is far away from the band gap, and we are eager to discard them. An elegant way to do this is to ignore some of the terms in the perturbation and consider the "rotating wave" Hamiltonian

$$H_{em}^{(rw)} = \frac{1}{2}\hbar\omega_R(t)e^{\imath\omega_0 t}\sum_{\mathbf{k},\sigma} a_{e,\sigma,\mathbf{k}}a_{h,-\sigma,-\mathbf{k}} + h.c..$$

The total Hamiltonian we start with will be

$$H(t) = H_{eh}^{HF}(t) + H_{em}^{(rw)},$$

where the electron–hole part is taken in the Hartree–Fock approximation

$$
\begin{aligned}
H_{eh}^{HF}(t) = \textstyle\sum_\sigma \sum_{\mathbf{k}} & \left((\frac{\hbar^2 k^2}{2m_e} + \frac{1}{2}E_g) a^+_{e,\sigma,\mathbf{k}} a_{e,\sigma,\mathbf{k}} + (\frac{\hbar^2 k^2}{2m_h} + \frac{1}{2}E_g) a^+_{h,\sigma,\mathbf{k}} a_{h,\sigma,\mathbf{k}} \right) \\
+ & \sum_{\sigma,\sigma'} \sum_{\mathbf{k},\mathbf{k}',\mathbf{q}} \frac{4\pi e^2}{q^2 V} \Big[a^+_{e,\sigma,\mathbf{k+q}} a_{e,\sigma,\mathbf{k}} \langle a^+_{e,\sigma',\mathbf{k}'-\mathbf{q}} a_{e,\sigma',\mathbf{k}'} \rangle_t \\
- & \; a^+_{e,\sigma,\mathbf{k+q}} a_{e,\sigma'\mathbf{k}'} \langle a^+_{e,\sigma',\mathbf{k}'-\mathbf{q}} a_{e,\sigma,\mathbf{k}} \rangle_t \\
+ & \; a^+_{h,\sigma,\mathbf{k+q}} a_{h,\sigma\mathbf{k}} \langle a^+_{h,\sigma',\mathbf{k}'-\mathbf{q}} a_{h,\sigma',\mathbf{k}'} \rangle_t \\
- & \; a^+_{h,\sigma,\mathbf{k+q}} a_{h,\sigma'\mathbf{k}'} \langle a^+_{h,\sigma',\mathbf{k}'-\mathbf{q}} a_{h,\sigma,\mathbf{k}} \rangle_t \\
- & \; a^+_{e,\sigma,\mathbf{k+q}} a_{e,\sigma\mathbf{k}} \langle a^+_{h,\sigma',\mathbf{k}'-\mathbf{q}} a_{h,\sigma',\mathbf{k}'} \rangle_t \\
- & \; a^+_{h,\sigma,\mathbf{k+q}} a_{h,\sigma\mathbf{k}} \langle a^+_{e,\sigma',\mathbf{k}'-\mathbf{q}} a_{e,\sigma',\mathbf{k}'} \rangle \\
- & \; a^+_{e,\sigma,\mathbf{k+q}} a^+_{h,\sigma',\mathbf{k}'-\mathbf{q}} \langle a_{h,\sigma',\mathbf{k}'} a_{e,\sigma,\mathbf{k}} \rangle_t \\
- & \; a_{h,\sigma,\mathbf{k}'} a_{e,\sigma',\mathbf{k}} \langle a^+_{e,\sigma',\mathbf{k+q}} a^+_{h,\sigma,\mathbf{k}'-\mathbf{q}} \rangle_t \Big] .
\end{aligned}
$$

Here we took into account all the anomalous averages that will be induced only by the electromagnetic field.
We may take into account also momentum conservation, spin independence as well as the charge neutrality of the total system. Then

$$
\begin{aligned}
\langle a^+_{e,\sigma',\mathbf{k}'} a_{e,\sigma,\mathbf{k}} \rangle_t &= \delta_{\mathbf{k},\mathbf{k}'} \delta_{\sigma,\sigma'} f_{e,\mathbf{k}}(t) \\
\langle a^+_{h,\sigma',\mathbf{k}'} a_{h,\sigma,\mathbf{k}} \rangle_t &= \delta_{\mathbf{k},\mathbf{k}'} \delta_{\sigma,\sigma'} f_{h,\mathbf{k}}(t) \\
\langle a_{e,\sigma',\mathbf{k}'} a_{h,\sigma,\mathbf{k}} \rangle_t &= \delta_{\mathbf{k},-\mathbf{k}'} \delta_{\sigma,-\sigma'} p_{\mathbf{k}}(t) \\
\sum_{\mathbf{k}} f_{e,\mathbf{k}}(t) &= \sum_{\mathbf{k}} f_{h.\mathbf{k}}(t)
\end{aligned}
$$

and the Hartree terms disappear. Then the following equations of motion emerge for the electron–hole populations $f_{e,\mathbf{k}}(t),\ f_{h,\mathbf{k}}(t)$ and the "inter-band polarization" $p_{\mathbf{k}}$

$$
\begin{aligned}
\frac{\partial}{\partial t} f_{e,\mathbf{k}}(t) &= \Im \left\{ \Omega^*_{\mathbf{k}}(t)\, p_{\mathbf{k}}(t) e^{\imath\omega_0 t} \right\} \\
\frac{\partial}{\partial t} f_{h,\mathbf{k}}(t) &= \Im \left\{ \Omega^*_{\mathbf{k}}(t)\, p_{\mathbf{k}}(t) e^{\imath\omega_0 t} \right\}
\end{aligned}
$$

$$
\frac{\partial}{\partial t} p_{\mathbf{k}}(t) + \frac{\imath}{\hbar} \left(\epsilon_{e,\mathbf{k}}(t) + \epsilon_{h,\mathbf{k}}(t) \right) p_{\mathbf{k}}(t) = \frac{\imath}{2} \Omega_k(t) e^{-\imath\omega_0 t} \left(1 - f_{e,\mathbf{k}}(t) - f_{h,\mathbf{k}}(t) \right) .
$$

Here $\epsilon_{e,\mathbf{k}}(t)$ and $\epsilon_{h,\mathbf{k}}(t)$ are the "renormalized" electron and hole energies

$$\epsilon_{e,\mathbf{k}}(t) \equiv \frac{\hbar^2 k^2}{2m_e} + \frac{1}{2}E_g + \frac{1}{V}\sum_{\mathbf{k}'\neq\mathbf{k}} V_{|\mathbf{k}-\mathbf{k}'|}\, f_{e,\mathbf{k}'}(t)$$

$$\epsilon_{h,\mathbf{k}}(t) \equiv \frac{\hbar^2 k^2}{2m_h} + \frac{1}{2}E_g + \frac{1}{V}\sum_{\mathbf{k}'\neq\mathbf{k}} V_{|\mathbf{k}-\mathbf{k}'|}\, f_{h,\mathbf{k}'}(t)$$

while $\Omega_{\mathbf{k}}(t)$ is the generalized Rabi frequency

$$\Omega_{\mathbf{k}}(t) \equiv \omega_R(t) + \frac{2}{\hbar}\frac{1}{V}\sum_{\mathbf{k}'\neq\mathbf{k}} V_{|\mathbf{k}-\mathbf{k}'|}\, p_{\mathbf{k}'}(t)e^{\imath\omega_0 t}\ ;\ V_q \equiv \frac{4\pi e^2}{q^2}.$$

The rapidly oscillating factor $e^{-\imath\omega_0 t}$ may be eliminated from the equation by a redefinition of the inter-band polarization

$$p_{\mathbf{k}}(t) \equiv \bar{p}_{\mathbf{k}}(t)e^{-\imath\omega_0 t}$$

and we obtain the "semiconductor Bloch equations" for the slowly varying entities $f_{e,\mathbf{k}}, f_{h,\mathbf{k}}, \bar{p}_{\mathbf{k}}$

$$\frac{\partial}{\partial t} f_{e,\mathbf{k}} = \Im\left\{\Omega_{\mathbf{k}}^*\, \bar{p}_{\mathbf{k}}\right\}$$

$$\frac{\partial}{\partial t} f_{h,\mathbf{k}} = -\Im\left\{\Omega_{\mathbf{k}}^*\, \bar{p}_{\mathbf{k}}\right\}$$

$$\left(\frac{\partial}{\partial t} + \frac{\imath}{\hbar}\Delta_{\mathbf{k}}\right)\bar{p}_{\mathbf{k}} = \frac{\imath}{2}\Omega_k\left(1 - f_{e,\mathbf{k}} - f_{h,\mathbf{k}}\right) - \frac{\bar{p}_{\mathbf{k}}}{T_2}$$

with the detuning

$$\Delta_{\mathbf{k}} \equiv \epsilon_{e,\mathbf{k}} + \epsilon_{h,\mathbf{k}} - \hbar\omega_0.$$

We have introduced here by hand also a phenomenological term with a "relaxation time" T_2 in order to take into account symbolically other interactions. This is a standard approach in the treatment of the non-linear optics of two level atoms in the frame of the Bloch equations on a quite different time scale. In semiconductors, the chosen time scale is such that the here ignored processes are still coherent and only on a longer time scale they destroy the coherence. The ultra-short-time experiments are performed precisely to study the kinetics of the electron–hole plasma on this time scale, where coherent quantum mechanical features play still an essential role. The many-body treatment, of these rapid processes, taking into account Coulomb interactions as well as the interaction with LO-phonons is called "quantum kinetics." This theory is rather complicated and we do not touch it.

Since the populations of electrons and holes are equal $f_e = f_h \equiv f$, one of these equations is superfluous. The above equations may be solved either numerically or by a systematic development of the solution in powers of the field, i.e., of $\omega_R(t)$.

6.8 Third Order Non-Linear Response

We outline here the scheme of the third order non-linear response. Since different **k** values in the semiconductor Bloch equations are not coupled we may omit the **k** index. Further, we simplify the discussion by omitting here the Hartree–Fock terms. The solution we look for is a systematic development in the powers of the electromagnetic perturbation

$$f(t) = f^{(1)}(t) + f^{(2)}(t) + f^{(3)}(t) \dots$$
$$\bar{p}(t) = \bar{p}^{(1)}(t) + \bar{p}^{(2)}(t) + \bar{p}^{(3)}(t) + \dots .$$

To first order we have

$$f^{(1)}(t) = 0$$
$$\bar{p}^{(1)}(t) = \frac{\imath}{2} \int_{-\infty}^{t} dt' \, e^{-(\frac{\imath}{\hbar}\Delta + \frac{1}{T_2})(t-t')} \, \omega_R(t'),$$

while to third order

$$f^{(3)}(t) = 0$$
$$\bar{p}^{(3)}(t) = -\imath \int_{-\infty}^{t} dt' \, e^{-(\frac{\imath}{\hbar}\Delta + \frac{1}{T_2})(t-t')} \, \omega_R(t') f^{(2)}(t').$$

Inserting in the last step the result of the previous approximation we get the only non-vanishing terms

$$f^{(2)}(t) = \frac{1}{2} \int_{-\infty}^{t} dt' \int_{-\infty}^{t'} dt'' \, \Re \left(\omega_R^*(t') \, \omega_R(t'') \, e^{-(\frac{\imath}{\hbar}\Delta + \frac{1}{T_2})(t'-t'')} \right)$$

and

$$\bar{p}^{(3)}(t) = -\frac{\imath}{2} \int_{-\infty}^{t} dt' \int_{-\infty}^{t'} dt'' \int_{-\infty}^{t''} dt''' e^{-(\frac{\imath}{\hbar}\Delta + \frac{1}{T_2})(t-t')} \, \omega_R(t')$$
$$\times \, \Re \left(\omega_R^*(t'') \, \omega_R(t''') \, e^{-(\frac{\imath}{\hbar}\Delta + \frac{1}{T_2})(t''-t''')} \right).$$

In what follows we shall consider that the optical field (laser) consists of a stronger pump pulse and a weaker, much shorter test pulse, both having the same carrier frequency ω_0, and also

$$\omega_R(t) = \omega_R^P(t) + \omega_R^T(t).$$

Since the test pulse is assumed to be very weak, we will retain only terms of first order in $\omega_R^T(t)$. (This restriction is, however, not compulsory in the analysis of the four wave mixing.) Then

$$\begin{aligned}\bar{p}^{(3)}(t) \approx & -\frac{\imath}{2}\int_{-\infty}^{t} dt' \int_{-\infty}^{t'} dt'' \int_{-\infty}^{t''} dt''' \; e^{-(\frac{\imath}{\hbar}\Delta + \frac{1}{T_2})(t-t')} \; \omega_R^T(t') \\ & \times \Re\left(\omega_R^{P*}(t'')\,\omega_R^P(t''')\,e^{-(\frac{\imath}{\hbar}\Delta + \frac{1}{T_2})(t''-t''')}\right) \\ & -\frac{\imath}{2}\int_{-\infty}^{t} dt' \int_{-\infty}^{t'} dt'' \int_{-\infty}^{t''} dt''' \; e^{-(\frac{\imath}{\hbar}\Delta + \frac{1}{T_2})(t-t')} \; \omega_R^P(t') \\ & \times \Re\left[\left(\omega_R^{P*}(t'')\,\omega_R^T(t''') + \omega_R^{T*}(t'')\,\omega_R^P(t''')\right) e^{-(\frac{\imath}{\hbar}\Delta + \frac{1}{T_2})(t''-t''')}\right].\end{aligned}$$

As one may see, there are two different structures, either

$$\bar{p} \propto \omega_R^T \cdot |\omega_R^P|^2$$

or

$$\bar{p} \propto \omega_R^{T*} \cdot (\omega_R^P)^2.$$

To interpret the significance of these structures we must return to the ignored propagation properties of the laser beams given by the phase factors $e^{\imath \mathbf{k}_T \mathbf{r}}$, respectively, $e^{\imath \mathbf{k}_P \mathbf{r}}$. While by the interaction within the semiconductor, these may be neglected; now their importance is very important in analyzing non-linear experiments. Since the excited inter-band polarization is itself a source of emerging electromagnetic waves, we expect that the first structure will cause an emerging beam in the $\mathbf{k}_T$ -direction, while the second in the $(2\mathbf{k}_P - \mathbf{k}_T)$-direction. The first is responsible for the differential transmission (DTS), while the second for the four wave mixing (FWM). The configuration of a DTS experiment is shown in Fig. 6.2, while that of an FWM experiment in Fig. 6.3. Here the pink arrows show the test and pump propagation. The pulses themselves are shown in violet, while the new signal in the $(2\mathbf{k}_P - \mathbf{k}_T)$ direction is shown with a red arrow.

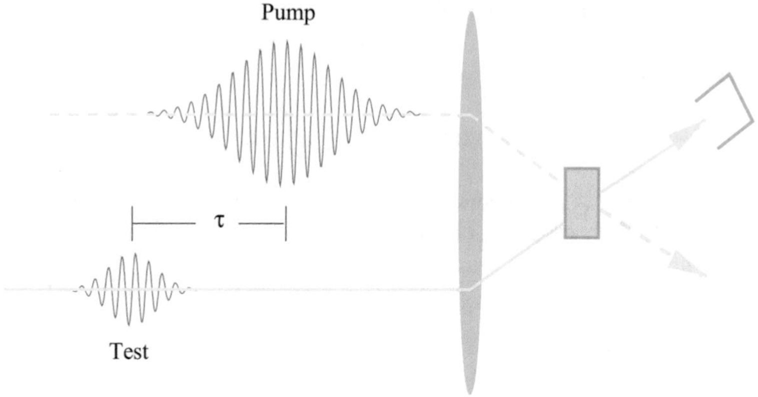

Fig. 6.2 Differential transmission experiment

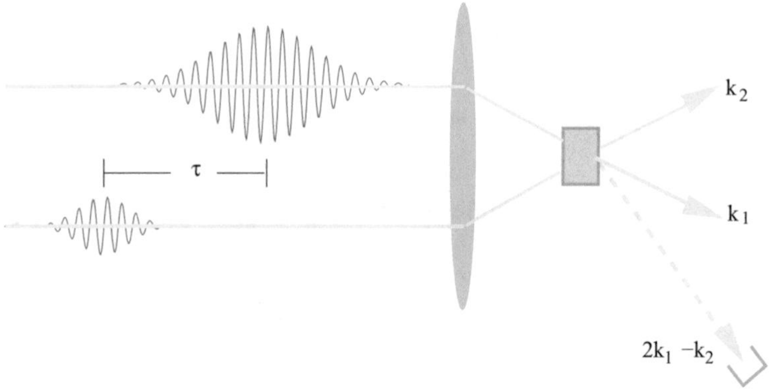

Fig. 6.3 Four wave mixing experiment

6.9 Differential Transmission

In this experiment, one uses two laser pulses of different widths with a retardation time τ between them. By the stronger pump pulse electron–hole pairs are excited, and the test pulse sees a prepared state, different from that in the absence of the pump. To identify the transmitted test pulse from the pump pulse one chooses a small angle between the pulses. By varying the retardation time τ one may obtain information about the evolution of the interacting electron–hole pairs created by the pump pulse.

Within the χ_3 theory outlined before we have

$$\bar{p}_{\mathbf{k}}^{(3)DTS}(t) = -\frac{i}{4}\int_{-\infty}^{t} dt' \int_{-\infty}^{t'} dt'' \int_{-\infty}^{t''} dt''' \; e^{-(\frac{i}{\hbar}\Delta_{\mathbf{k}}+\frac{1}{T_2})(t-t')} \; \omega_R^T(t')$$

$$\times \left(\omega_R^{P*}(t'')\, \omega_R^{P}(t''')\, e^{(-\frac{\imath}{\hbar}\Delta_{\mathbf{k}}-\frac{1}{T_2})(t''-t''')} + \omega_R^{P}(t'')\, \omega_R^{P*}(t''')\, e^{(\frac{\imath}{\hbar}\Delta_{\mathbf{k}}-\frac{1}{T_2})(t''-t''')} \right)$$

$$- \frac{\imath}{4} \int_{-\infty}^{t} dt' \int_{-\infty}^{t'} dt'' \int_{-\infty}^{t''} dt'''\; e^{-(\frac{\imath}{\hbar}\Delta_{\mathbf{k}}+\frac{1}{T_2})(t-t')}\, \omega_P(t') \times \left[\omega_R^{P*}(t'')\, \omega_R^{T}(t''')\, e^{-(\frac{\imath}{\hbar}\Delta_{\mathbf{k}}+\frac{1}{T_2})(t''-t''')} \right].$$

If the pulses do not overlap one may retain only the term

$$\bar{p}^{(3)DTS}(t) \approx -\frac{\imath}{8} \int_{-\infty}^{t} dt'\; e^{-(\frac{\imath}{\hbar}\Delta+\frac{1}{T_2})(t-t')}\, \omega_R^{T}(t') \left| \int_{-\infty}^{t'} dt'' \omega_R^{P}(t'') e^{\frac{\imath}{\hbar}\Delta t''} \right|^2 .$$

If the test pulse is very short, one gets further

$$\bar{p}^{(3)DTS}(t) \approx -\frac{\imath}{8} e^{-(\frac{\imath}{\hbar}\Delta_{\mathbf{k}}+\frac{1}{T_2})(t-\tau)} \theta(t-\tau) \int_{-\infty}^{\infty} dt' \omega_R^{T}(t') \left| \int_{-\infty}^{\tau} dt'' \omega_R^{P}(t'') e^{\frac{\imath}{\hbar}\Delta_{\mathbf{k}} t''} \right|^2$$

or

$$\bar{p}^{(3)DTS}(t) \approx -\frac{\imath}{8} \theta(t-\tau) e^{-(\frac{\imath}{\hbar}\Delta+\frac{1}{T_2})(t-\tau)} \tilde{\omega}_R^{T}(0) |\tilde{\omega}_R^{P}(\Delta)|^2,$$

where $\tilde{\omega}_R^P(\Delta)$ and $\tilde{\omega}_R^T(0)$ are the Fourier transformed pulse and test Rabi frequencies at the detuning Δ, respectively, at $t = 0$.

The total DTS polarization $P^{(3)DTS}(t)$ results after summation over all the wave vectors and re-multiplication with $e^{-\imath\omega_0 t}$. The Fourier analysis of the resulting time-dependent signal shows the modification of the test pulse absorption due to the presence of the pump pulse. This DTS signal (differential transmission) is defined as the difference between the absorption of the test pulse in the presence of the pump pulse and that in the absence of the pump. Now, it may be shown further that this difference reaches its maximal negative value again at the carrier frequency ω_0, if the test pump retardation time τ is much bigger than the width of the pulses. The meaning of this result is that one cannot excite electron–hole pairs in a domain that already was emptied by the pump. This phenomenon is called "hole burning." However, during the time τ the electron–hole system, due to the Coulomb interaction as well as the interaction with phonons, changes its state. By measuring the DTS signal at different retardations τ one may get information about this evolution and compare it with theoretical predictions. Since in this simple description we did not include such an evolution, we do not follow further the explicit calculus.

6.10 Four Wave Mixing

The FWM signal within the χ_3 theory is contained in the terms proportional to $\omega_R^{T*}(\omega_R^P)^2$

$$\bar{p}^{(3)VWM}(t) = -\frac{\iota}{4}\int_{-\infty}^{t} dt' \int_{-\infty}^{t'} dt'' \int_{-\infty}^{t''} dt''' \; e^{-(\frac{\iota}{\hbar}\Delta+\frac{1}{T_2})(t-t')}\,\omega_R^P(t')$$
$$\times \left(\omega_R^{T*}(t'')\,\omega_R^P(t''')\,e^{-(\frac{\iota}{\hbar}\Delta+\frac{1}{T_2})(t''-t''')}\right.$$
$$\left. + \;\omega_R^P(t'')\,\omega_R^{T*}(t''')\,e^{(\frac{\iota}{\hbar}\Delta-\frac{1}{T_2})(t''-t''')}\right).$$

Unlike in the case of the DTS experiment, one chooses here two pulses with a large retardation time τ. Under this condition only the second term of the previous equation contributes

$$\bar{p}^{(3)VWM}(t) = -\frac{\iota}{4}\int_{-\infty}^{t} dt' \int_{-\infty}^{t'} dt'' \int_{-\infty}^{t''} dt'''$$
$$\times \; e^{\frac{\iota}{\hbar}\Delta(t'-t+t''-t''')} e^{-\frac{(t-t'+t''-t''')}{T_2}}\;\omega_R^P(t')\omega_R^P(t'')\omega_R^T(t''').$$

If one considers that the two pulses do not overlap at all, one gets

$$P^{(3)VWM}(t) = -\frac{\iota}{8}e^{-\frac{t}{T_2}}\tilde{\omega}_R^P(0)^2\tilde{\omega}_R^T(0)\int d\mathbf{k}e^{\frac{\iota}{\hbar}\Delta_{\mathbf{k}}(t-2\tau)}.$$

The integral over **k** diverges and one must restrict it to the Brillouin zone. Due to the oscillating factor, the integral reaches its maximum at $t = 2\tau$ and therefore one may observe a “photon echo” in the $2\mathbf{k}_P - \mathbf{k}_T$-direction after a delay time 2τ, its height decreasing as $e^{-\frac{2\tau}{T_2}}$.

The outlined theory of these experiments is oversimplified, illustrating only the essential idea. The deviations from the simple one-particle treatment due to the Coulomb interaction and phonon emission/absorption are the main information one follows, and the height of the photon echo is not given by the above simple exponential decay. Actually, this kind of optical experiments (due to the retardation time parameter) delivers a sort of motion picture of the evolution of the electron–hole plasma.

Chapter 7
Phase Transitions

A few examples of phase transitions allow an insight into the fine aspects of the thermodynamic limit. Spontaneous symmetry breaking at critical temperatures or densities leads to unexpected stable states characterized by order parameters. We describe at the beginning the ferromagnetic transition within the frame of the Heisenberg model of localized spins. The mathematical subtlety of the Bose–Einstein condensation is discussed in some more detail, including its description in real time. The excitation spectrum of the s.c. Bogoliubov model of repulsive massive bosons at $T = 0$ is discussed. Its time-dependent extension and the Gross–Pitaevskii equation with a simple illustration are also included. We devote more place to superconductivity by describing first the phenomenological theory of London, followed by a quantum mechanical treatment in real space, within the frame of a simple model of an effective interaction between the electrons suggested by the BCS theory.

An important aspect of the physics of condensed matter is the possibility of phase transitions. Their theoretical treatment is essentially based on the thermodynamic limit. We do not intend to describe the multitude of the phase transitions, but just want to take a glimpse at the basic ideas within some specific models. Some of the most spectacular new phenomena although subject of abundant literature are still not ripe for a simple presentation.

L. A. Bányai, *A Compendium of Solid State Theory*,
https://doi.org/10.1007/978-3-030-37359-7_7

7.1 The Heisenberg Model of Ferro-Magnetism

Ferro-magnetism is thought to originate from the spin-dependent exchange interaction of the core electrons of neighboring ions. A simple model is due to Heisenberg that considers N interacting spins $\mathbf{S}$ on a periodic lattice $\mathbf{R}$. The Hamiltonian in the presence of a magnetic field $\mathbf{B} = (0, 0, B)$ oriented along the z axis is

$$H = -\frac{1}{2} \sum_{\mathbf{R}\mathbf{R}'} J(\mathbf{R} - \mathbf{R}')\mathbf{S}_{\mathbf{R}}\mathbf{S}_{\mathbf{R}'} - 2\mu_0 B \sum_{\mathbf{R}} S^z_{\mathbf{R}},$$

where μ_0 is the magneton and the spin-operators $\mathbf{S}$ obey the commutation rules

$$\left[S^+_{\mathbf{R}}, S^-_{\mathbf{R}}\right] = 2S^z_{\mathbf{R}}$$
$$\left[S^z_{\mathbf{R}}, S^\pm_{\mathbf{R}}\right] = \pm S^z_{\mathbf{R}}$$

with $S^\pm_{\mathbf{R}} = S^x_{\mathbf{R}} \pm S^y_{\mathbf{R}}$. The commutators of spin-operators on different sites vanish. The coupling $J(\mathbf{R}) \geq 0$ is supposed to vanish rapidly with the distance R. One may write the Hamiltonian also in the form

$$H = -\frac{1}{2} \sum_{\mathbf{R},\mathbf{R}'} J(\mathbf{R} - \mathbf{R}') \left(S^z_{\mathbf{R}} S^z_{\mathbf{R}'} + \frac{S^+_{\mathbf{R}} S^-_{\mathbf{R}'} + S^-_{\mathbf{R}} S^+_{\mathbf{R}'}}{2} \right) - 2\mu_0 B \sum_{\mathbf{R}} S^z_{\mathbf{R}}$$

to realize that the total spin $\mathbf{S}_{tot}$ in the z-direction

$$S^z_{tot} \equiv \sum_{\mathbf{R}} S^z_{\mathbf{R}}$$

is conserved

$$\left[S^z_{tot}, H\right] = 0.$$

Therefore, the energy eigenstates are also eigenstates of S^z_{tot}, as well as of $\mathbf{S_{tot}}^2$ due to rotation invariance. A simple basis in the spin space may be built up from products of eigenstates of each spin $|S, m\rangle$with m being its projection on the z axis. The state with the maximum of total spin is the one with all the spins aligned in the z-direction

$$|\Phi_0\rangle = |S, S\rangle_{\mathbf{R}_1} |S, S\rangle_{\mathbf{R}_2} \dots |S, S\rangle_{\mathbf{R}_N}.$$

This state is an eigenstate of H

$$H|\Phi_0>=\left(-\frac{1}{2}S^2\sum_{\mathbf{R},\mathbf{R}'}J(\mathbf{R}-\mathbf{R}')-2\mu_0BNS\right)|\Phi>$$

since

$$S^+_{\mathbf{R}}|S,S\rangle_{\mathbf{R}}=0.$$

Consider now the average of the Hamiltonian over an arbitrary state $|\Phi\rangle$ of our basis

$$|\Phi\rangle=|S,m_1\rangle_1|S,m_2\rangle_2\ldots|S,m_N\rangle_N.$$

Since the terms non-commuting with $S^z_{\mathbf{R}}$vanish in the average, one gets

$$\langle\Phi|H|\Phi\rangle=-\frac{1}{2}\sum_{\mathbf{R},\mathbf{R}'}J(\mathbf{R}-\mathbf{R}')m_{\mathbf{R}}m_{\mathbf{R}'}-2\mu_0B\sum_{\mathbf{R}}m_{\mathbf{R}}$$

and this is obviously greater than the eigenenergy of the state $|\Phi_0\rangle$. Therefore, $|\Phi_0\rangle$ is the ground state of the system with the ground state energy

$$E_0=-\frac{1}{2}S^2\sum_{\mathbf{R},\mathbf{R}'}J(\mathbf{R}-\mathbf{R}')-2\mu_0BNS.$$

In this state the total spin in the z direction is NS and the per node magnetic moment (magnetization) is

$$M=2\mu_0S.$$

In the absence of the magnetic field B the choice of the quantization axis is arbitrary, and this state would have had a $2NS+1$ fold degeneracy. The average over all these states of the same energy would give rise to no magnetization. Thus, with a finite number of spins the Heisenberg model would not lead to a stable ferromagnetic state in the absence of an external magnetic field. However, the situation changes drastically if one considers the thermodynamic limit. Indeed, one sees that by $N,\Omega\to\infty$ by $\frac{N}{\Omega}=n$ (constant spin density) even an infinitesimal magnetic field may determine the direction of the magnetization, while its magnitude is the same $M=2\mu_0S$. This is the *spontaneous symmetry breaking*.
Let us treat the thermodynamics of this model in the absence of a magnetic field **B** in the mean-field (self-consistent) approximation in which one of the spin-operators we replace by its average

$$H_{mf}=-\sum_{\mathbf{R}}2\mu_0\langle\mathbf{B}\rangle\mathbf{S}_{\mathbf{R}}$$

with the effective (s.c.) magnetic field being

$$\langle \mathbf{B} \rangle = \frac{1}{2\mu_0} \sum_{\mathbf{R}} J(\mathbf{R}) \langle \mathbf{S_R} \rangle$$

and we look for a solution with the average on site spin $\langle \mathbf{S_R} \rangle$ as being site independent and oriented along the z direction. Then

$$H_{mf} = -\sum_{\mathbf{R}} 2\mu_0 \mathscr{B} S^z_{\mathbf{R}}; \qquad \mathscr{B} = \frac{1}{2\mu_0} \langle S^z \rangle \sum_{\mathbf{R}} J(\mathbf{R}).$$

The statistical sum is

$$Z = Spe^{-\beta H_{mf}} = \left(\sum_{S^z} e^{-2\beta\mu_0 \mathscr{B} S^z} \right)^N,$$

while the free energy and the equilibrium magnetization are given by

$$F \equiv \frac{1}{N} \ln \left(\sum_{S^z} e^{-2\beta\mu_0 \mathscr{B} S^z} \right)$$

and

$$M \equiv -\frac{\partial F}{\partial \mathscr{B}} = 2\beta\mu_0 \langle S^z \rangle.$$

For spin $S = \frac{1}{2}$ one gets

$$M = \beta\mu_0 \tanh(\beta\mu_0 \mathscr{B})$$

that has to be considered together with the self-consistency equation

$$\mathscr{B} = \frac{1}{4\mu_0} \tanh(\beta\mu_0 \mathscr{B}) \sum_{\mathbf{R} \neq 0} J(\mathbf{R}).$$

Since $\frac{\tanh(x)}{x} < 1$, over the critical temperature

$$T_c = \frac{1}{4k_B} \sum_{\mathbf{R} \neq 0} J(\mathbf{R})$$

one has only a vanishing solution for the effective magnetic field $\mathscr{B}$.

However for $T < T_c$ there is a non-vanishing solution and consequently, a finite magnetization in the absence of any external magnetic field. At $T = T_c$ a ferromagnetic phase transition occurs.

7.2 Bose Condensation

Another good example of a phase transition is the Bose condensation. This is the only phase transition that occurs without any interaction. In condensed matter, such a transition may occur for the excitons as massive bosons. Let us consider a system of free massive bosons in a finite volume (with periodical boundary conditions) described by the second quantized Hamiltonian

$$H = \sum_{\mathbf{k}} e_{\mathbf{k}} a_{\mathbf{k}}^{+} a_{\mathbf{k}}$$

having the one-particle energies

$$e_{\mathbf{k}} = \frac{\hbar^2 \mathbf{k}^2}{2m}.$$

The macro-canonical statistical distribution gives the average number of bosons in the state of wave vector $\mathbf{k}$ to be the Bose distribution

$$\mathscr{N}(e_{\mathbf{k}}) \equiv \frac{1}{e^{\beta(e_{\mathbf{k}}-\mu)} - 1}. \tag{7.1}$$

Obviously, the chemical potential must be negative $\mu < 0$, otherwise the expression is negative below $k = 0$. The average total number of bosons fixes μ

$$\langle n \rangle = \frac{\langle N \rangle}{\Omega} = \frac{1}{\Omega} \sum_{\mathbf{k}} \frac{1}{e^{\beta(e_{\mathbf{k}}-\mu)} - 1},$$

with $\Omega = L^3$ and $\mathbf{k} = (\frac{2\pi n_x}{L}, \frac{2\pi n_y}{L}, \frac{2\pi n_z}{L})$, or in the thermodynamic limit

$$\langle n \rangle = \int \frac{d^3k}{(2\pi)^3} \frac{1}{e^{\beta(e_{\mathbf{k}}+\mu)} - 1}.$$

This 3D integral, however, reaches its possible maximum value at a critical density n_c for $\mu = 0$

$$n_c = \int \frac{d^3k}{(2\pi)^3} \frac{1}{e^{\beta e_{\mathbf{k}}} - 1} = \frac{1}{8\pi^{3/2}} \zeta\left(\frac{3}{2}\right) \frac{1}{\ell^3} = 0.0586436 \frac{1}{\ell^3}.$$

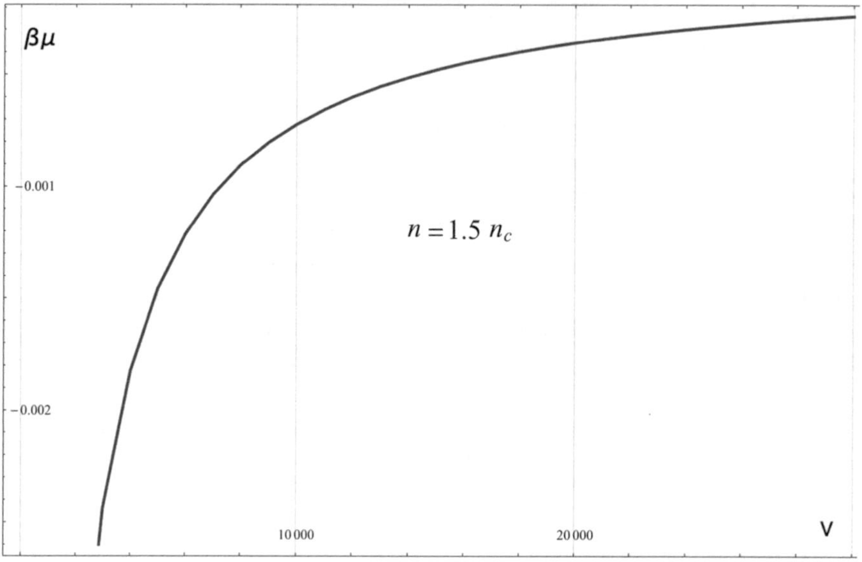

Fig. 7.1 The dependence of $\beta\mu$ on the volume $v = \frac{\Omega}{\ell^3}$ for $n = 1.5n_c$

Here we have introduced the natural length $\ell = \sqrt{\frac{\beta\hbar^2}{2m}}$ we shall use further in order to define also a dimensionless volume $v = \Omega/\ell^3$. The analogous integrals in 1D or 2D have no upper limits! Therefore it is a pure 3D feature. Seemingly the system cannot accommodate any higher average density! As one may see, the key of the paradox lies in the contribution of the lowest, $\mathbf{k} = 0$ state in the discrete sum

$$\frac{1}{\Omega} \frac{1}{e^{\beta\mu} - 1}.$$

So far the chemical potential is negative and finite; its contribution is infinitesimal as $\Omega \to \infty$ and it is a piece of the Riemann sum. However, as the chemical potential goes to 0, the value of the above ratio is determined by the way μ goes to 0 as function of the volume Ω.

As it is illustrated in Fig. 7.1, above the critical density n_c, as the volume Ω increases $\beta\mu$ goes to zero from negative values as

$$-\frac{1}{n_{cond}\Omega}$$

and therefore, the contribution of the $\mathbf{k} = 0$ state to the sum remains finite and equal to n_{cond}. This contribution has to be added thus independently.

Therefore, the correct equation for $n \geq n_c$ is

$$n_{cond} + \int \frac{d^3k}{(2\pi)^3} \frac{1}{e^{\beta e_k} - 1} = n.$$

Above the critical density the chemical potential remains zero but the condensate density n_{cond} increases.
The critical situation may be achieved either by increasing the density at a fixed temperature ($n \to n_c$, at fixed β), or by lowering the temperature at a fixed density ($\beta \to \beta_c$ at fixed density n).

Like the ferromagnetic phase transition, the Bose condensation shows a spontaneous symmetry breaking. This may be seen within the Bogoliubov approach. In this approach, one introduces an infinitesimal symmetry breaking term in the Hamiltonian before the thermodynamic limit and let it vanish after the limit is already performed.
Bogoliubov introduces terms that break invariance against a multiplication of the creation/annihilation operators by a phase factor, therefore particle number conservation is violated. In the macro-canonical statistical sum

$$Z = Tr\{e^{-\beta(H-\mu N)}\}$$

one considers

$$H - \mu N = \sum_{\mathbf{k}} (e_{\mathbf{k}} - \mu) a_{\mathbf{k}}^{+} a_{\mathbf{k}} + \lambda^* \sqrt{\Omega} a_0 + \lambda \sqrt{\Omega} a_0^{+}$$

with a small parameter λ. A simple c-number shift

$$A_0 = a_0 - \frac{\lambda\sqrt{\Omega}}{\mu}$$

defines the new bosonic annihilation operators for the state $\mathbf{k} = 0$ and helps to bring the expression again to a quadratic form

$$H - \mu N = \sum_{\mathbf{k} \neq 0} (e_{\mathbf{k}} - \mu) a_{\mathbf{k}}^{+} a_{\mathbf{k}} - \mu A_0^{+} A_0 + \frac{|\lambda|^2 \Omega}{\mu}.$$

The free energy is then

$$F = \frac{1}{\beta\Omega} \sum_{\mathbf{k}} \ln\{1 - e^{-\beta(e_{\mathbf{k}} - \mu)}\} + \frac{|\lambda|^2}{\mu}$$

leading to the average boson density

$$\langle n\rangle = \frac{\langle N\rangle}{\Omega} = \frac{1}{\Omega}\sum_{\mathbf{k}} \frac{1}{e^{\beta(e_{\mathbf{k}}-\mu)}-1} + \frac{|\lambda|^2}{\mu^2}.$$

Now we may apply without hesitating the limit rule for getting integrals out of sums, since the chemical potential remains always negative and get

$$\langle n\rangle = \frac{|\lambda|^2}{\mu^2} + \int \frac{d^3k}{(2\pi)^3} \frac{1}{e^{\beta(e_{\mathbf{k}}-\mu)}-1}.$$

At the critical density n_c the chemical potential vanishes in the $|\lambda| \to 0$ limit as

$$\mu \to \frac{|\lambda|}{\sqrt{n_{cond}}}$$

and we get the former result.

A new aspect of this approach is that here we explicitly recognize an order parameter λ that is the analog of the magnetization with the surviving phase of the symmetry breaking parameter ($\lambda = |\lambda| e^{\imath\phi}$) and the average value $\langle a_0\rangle$ of the original annihilation operator of the $\mathbf{k} = 0$ state becomes macroscopic (proportional to the square root of the total number of condensate bosons $N_{cond} = n_{cond}\Omega$)

$$\langle a_0\rangle = e^{\imath\phi}\sqrt{N_{cond}}.$$

A characteristic property of the condensed state is the existence of long range space correlations. The Riemann sum in the correlation function

$$\langle \psi(\mathbf{x})\psi^+(\mathbf{y})\rangle \equiv \frac{1}{\Omega}\sum_{\mathbf{k}} e^{\imath\mathbf{k}(\mathbf{x}-\mathbf{y})}\langle a_{\mathbf{k}}^+ a_{\mathbf{k}}\rangle = \frac{1}{\Omega}\sum_{\mathbf{k}} e^{\imath\mathbf{k}(\mathbf{x}-\mathbf{y})}\frac{1}{e^{\beta(e_{\mathbf{k}}-\mu)}-1}$$

above the critical temperature T_c goes over into the integral

$$\int \frac{d^3k}{(2\pi)^3}\frac{e^{\imath\mathbf{k}(\mathbf{x}-\mathbf{y})}}{e^{\beta e_{\mathbf{k}}}-1}$$

that vanishes as a power law for $|\mathbf{x}-\mathbf{y}| \to \infty$, while below the critical temperature one gets

$$\langle \psi(\mathbf{x})\psi^+(\mathbf{y})\rangle = n_{cond} + \int \frac{d^3k}{(2\pi)^3}\frac{e^{\imath\mathbf{k}(\mathbf{x}-\mathbf{y})}}{e^{\beta e_{\mathbf{k}}}-1}$$

and there is a term surviving over infinite distances.

7.2.1 Bose Condensation in Real Time

It is instructive to understand the real time description of a phase transition. We illustrate it here on the example of the Bose condensation within the frame of a rate equation description. The rate equation for massive bosons of wave vector $\mathbf{k}$ and energy $\epsilon(\mathbf{k}) = \frac{\hbar^2 k^2}{2m}$ interacting with a thermostat looks as

$$\frac{\partial}{\partial t}\langle n_{\mathbf{k}}(t)\rangle = -\sum_{\mathbf{k}'}\left[w_{\mathbf{k}\mathbf{k}'}\langle n_{\mathbf{k}}(t)\rangle\left(1+\langle n_{\mathbf{k}'}(t)\rangle\right) - w_{\mathbf{k}'\mathbf{k}}\langle n_{\mathbf{k}'}(t)\rangle\left(1+\langle n_{\mathbf{k}}(t)\rangle\right)\right]. \tag{7.2}$$

The transition rates $w_{\mathbf{k}\mathbf{k}'}$ are assumed to obey the detailed balance relation

$$w_{\mathbf{k}\mathbf{k}'} = e^{\beta(\epsilon(\mathbf{k})-\epsilon(\mathbf{k}'))} w_{\mathbf{k}'\mathbf{k}}.$$

Its derivation was left as a good exercise for the reader (see Sect. 12.3).

Obviously the stable equilibrium solution of this equation is the Bose distribution Eq. 7.1 for any $\mu < 0$. This equation conserves the total number of bosons. However, as we have shown before, as $\mu \to -0$ the density of bosons described by the Bose distribution with $\mu = 0$ cannot exceed a certain critical value n_c.

We have to modify the rate equation, in the spirit of the previous discussions, in order to take here also into account the accumulation of a macroscopic density of bosons in the state $\mathbf{k} = 0$. Therefore, we get a system of coupled rate equations for the average particle number in the "normal" states $f(\mathbf{k}, t) = \langle n_{\mathbf{k}}(t)\rangle$ and for $n_0(t) \equiv \frac{1}{\Omega}\langle n_0(t)\rangle$ with the volume of a cube $\Omega \equiv L^3$ and with cyclic boundary conditions $\Delta k = \frac{2\pi}{L}$, while taking into account that the transition rates are inversely proportional to the volume, i.e., $w_{\mathbf{k},\mathbf{k}'} = (\Delta k)^3 W(\mathbf{k}, \mathbf{k}')$. In the infinite volume limit ($L \to \infty$) one gets

$$\begin{aligned}\frac{\partial}{\partial t} f(\mathbf{k}, t) &= \int d\mathbf{k}\left\{W(\mathbf{k}, \mathbf{k}')f(\mathbf{k}, t)(1+f(\mathbf{k}', t)) - (\mathbf{k} \rightleftharpoons \mathbf{k}')\right\} \\ &\quad - \left[W(\mathbf{k}, 0)f(\mathbf{k}, t) - W(0, \mathbf{k})(1+f(\mathbf{k}, t))\right] n_0(t), \\ \frac{\partial}{\partial t} n_0(t) &= \int d\mathbf{k}\left[W(\mathbf{k}, 0)f(\mathbf{k}, t) - W(0, \mathbf{k})(1+f(\mathbf{k}, t))\right] n_0(t).\end{aligned} \tag{7.3}$$

With any small, but finite $n_0(0)$ this system of equations has a solution even above the critical density with increasing $n_0(t)$ to a finite value, while the normal density decreases correspondingly as to keep a constant total average density

$$\langle n\rangle = n_0(t) + \int d\mathbf{k} f(\mathbf{k}, t).$$

The non-linear equations 7.3 have two attractive fixed points for $t \to \infty$:

1. a "normal" one $f(\mathbf{k}) = \frac{1}{e^{\beta(e(\mathbf{k})-\mu)}-1}$ (with $\mu < 0$ and $n_0 = 0$), whose basin of attraction contains all the initial conditions with the total density below the critical one n_c and
2. a "condensed" one $f(\mathbf{k}) = \frac{1}{e^{\beta e(\mathbf{k})}-1}$ with $n_0 \neq 0$ whose basin of attraction contains all the initial conditions with the total density above the critical one n_c and simultaneously having any non-vanishing initial macroscopic condensate $n_0(0)$.

In order to produce BEC we need only an infinitesimal condensate "seed." This is a typical aspect of any phase transition in real time and represents the analog of Bogoliubov's approach to BEC.

In Figs. 7.2 and 7.3 we illustrate the evolution in time of the normal and condensate densities in time as described by Eq. 7.3 in two extreme cases. In the first figure one is below the critical total density, but starts with a strong condensate density and a very small non-condensate, while in the second figure the one starts with total density above the critical one starting with a strong non-condensate density and a very small condensate "seed." One may see that below critical density an initial condensate disappears, while above the total critical density it increases from a small "seed" to a finite value, while the normal density decreases correspondingly.

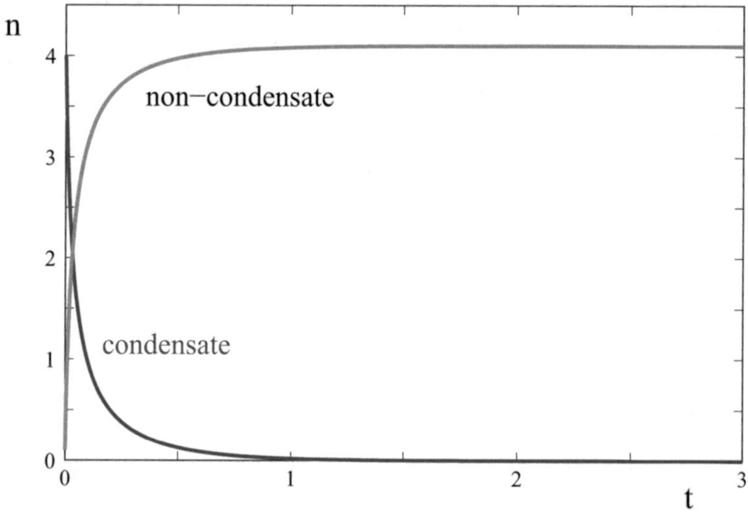

Fig. 7.2 Evolution of the non-condensate and condensate densities below the critical total density $n_c = 5.65$ starting with a strong condensate density and a very small non-condensate density (arbitrary units!)

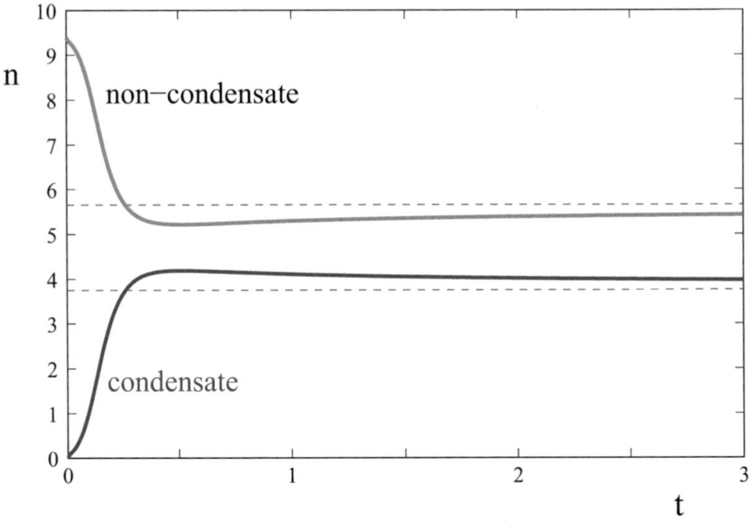

Fig. 7.3 Evolution of the non-condensate and condensate densities above the total critical density $n_c = 5.65$ starting with a very small condensate density and a finite non-condensate density.(Arbitrary units!)

7.3 Bogoliubov's Self-Consistent Model of Repulsive Bosons at T = 0

We shall discuss here the condensation of interacting bosons. The simplest version is a repulsive contact interaction described by the Hamiltonian

$$H = \int d\mathbf{x} \left\{ \psi(\mathbf{x})^+ \left(-\frac{\hbar^2}{2m} \nabla^2 + U(\mathbf{x}) \right) \psi(\mathbf{x}) + \frac{w}{2} \psi(\mathbf{x})^+ \psi(\mathbf{x})^+ \psi(\mathbf{x}) \psi(\mathbf{x}) \right\} .$$

Of course one has to still resort to approximations in order to tackle this many-body problem.

Bogoliubov's approach to this problem is a self-consistent Hamiltonian with the self-consistent Hartree term $\langle \psi(\mathbf{x})^+ \psi(\mathbf{x}) \rangle$ and a symmetry breaking one $\langle \psi(\mathbf{x}) \rangle$. The self-consistency is imposed on the ground state of the system ($T = 0$). Higher order symmetry breaking averages are considered implicitly as reducible to the lower order one, i.e., $\langle \psi(\mathbf{x}) \psi(\mathbf{x}) \rangle = \langle \psi(\mathbf{x}) \rangle^2$. In this frame one cannot discuss the phase transition, but just the self-consistency conditions for the assumed symmetry breaking phase. The Bogoliubov Hamiltonian is

$$
\begin{aligned}
H_B = \int d\mathbf{x} \Bigg\{ & \psi(\mathbf{x})^+ \left(-\frac{\hbar^2}{2m}\nabla^2 + \right) \psi(\mathbf{x}) \\
& + \frac{w}{2} \Big[4\psi(\mathbf{x})^+ \psi(\mathbf{x}) \langle \psi(\mathbf{x})^+ \psi(\mathbf{x}) \rangle \\
& + \left(\psi^+(\mathbf{x})\psi^+(\mathbf{x}) \langle \psi(\mathbf{x}) \rangle^2 + h.c. \right) \\
& - \left(4\psi(\mathbf{x}) \langle \psi(\mathbf{x})^+ \rangle |\langle \psi(\mathbf{x}) \rangle|^2 + h.c. \right) \Big] \Bigg\} + C_{HF}.
\end{aligned} \tag{7.4}
$$

The HF constant C_{HF} may be chosen as to satisfy the requirement

$$
\langle H \rangle = \langle H_B \rangle
$$

within the same sort of s.c. approximations. In the following discussion it plays no role and we omit it.

Rewriting the Hamiltonian in the **k**-space by

$$
\psi(\mathbf{x}) = \frac{1}{\sqrt{V}} \sum_{\mathbf{k}} b_{\mathbf{k}} e^{\imath \mathbf{k}\mathbf{x}}
$$

we get

$$
H_B = \sum_k \{ e_k b_k^+ b_k + \frac{1}{2}(c^* b_k b_{-k} + c b_{-k}^+ b_k^+) \} - 2\frac{w}{V} |\langle b_0 \rangle|^2 (\langle b_0 \rangle^* b_0 + \langle b_0 \rangle b_0^+), \tag{7.5}
$$

where we omitted the vector notation and used the notations

$$
e_k \equiv \frac{\hbar^2 k^2}{2m} + 2\frac{w}{V}\langle N \rangle = \frac{\hbar^2 k^2}{2m} + 2wn; \qquad c = w\frac{1}{V}\langle b_o \rangle^2 = w n_0.
$$

Taking into account the arbitrary phase in the definition of the creation–annihilation operators we have chosen here $\langle b_0 \rangle$ to be real and implicitly $c > 0$. In the case of a symmetry breaking phase transition it should survive in the thermodynamic limit.

For the case of particle non-conserving symmetry breaking, in the discussion of the equilibrium properties one should use not the Hamiltonian itself, but $\mathscr{H}_B \equiv H_B - \mu N$. One may diagonalize the $k \neq 0$ part of $\mathscr{H}_B$ by a canonical transformation to the new boson fields α_k and α_{-k}

$$
b_k = u_k \alpha_k + v_k \alpha_{-k}^+ \tag{7.6}
$$

with

$$
u_k = u_{-k}, \quad v_k = v_{-k} \quad \text{and} \quad |u_k|^2 - |v_k|^2 = 1\,.
$$

One gets

$$u_k = \cosh(\chi_k)e^{\imath\phi_k}; \; v_k = -\sinh(\chi_k)e^{-\imath\phi_k} \tag{7.7}$$

with

$$\tanh(2\chi_k) = \frac{c}{e_k - \mu}. \tag{7.8}$$

One has of course

$$\frac{c}{|e_k - \mu|} = \frac{wn_0}{\frac{\hbar^2 k^2}{2m} + 2wn} < 1.$$

The phase ϕ_k remains arbitrary and the energies are given by

$$\epsilon_k = sign(e_k - \mu)\sqrt{(e_k - \mu)^2 - c^2}.$$

For reasons of stability one should have $e_k - \mu > 0$ (for any k) and therefore $\mu < 2wn$. Thus

$$\epsilon_k = \sqrt{(e_k - \mu)^2 - c^2} = \sqrt{(\frac{\hbar^2 k^2}{2m} + 2wn - \mu)^2 - (wn_0)^2}.$$

Now we must proceed to discuss the self-consistency requirements stemming from the fact that the system is in its ground state. Then for $k \neq 0$

$$\langle \alpha_k^+ \alpha_k \rangle = 0 \qquad \text{and} \qquad \langle \alpha_k \rangle = 0$$

implying also

$$\langle b_k \rangle = 0$$

and according to the already mentioned assumption $\langle \alpha_k \alpha_{-k} \rangle = \langle \alpha_k \rangle \langle \alpha_{-k} \rangle$ it follows that

$$\langle b_k^+ b_k \rangle = 0,$$

i.e., what one could already expect in the ground state,

$$n = n_0, \tag{7.9}$$

i.e., the condensation is complete.

We have to discuss also the part of $\mathscr{H}_B$ containing only the operators of the condensate b_0:

$$(e_0 - \mu)b_0^+ b_0 + \frac{1}{2}c(b_0^+ b_0^+ + b_0 b_0) - 2w\sqrt{V}n_0^{\frac{1}{3}}(b_0 + b_0^+).$$

Before its diagonalization we must perform an allowed shift of the operators b_0 with a real constant *const*

$$b_0 \to b_0 + const$$

in order to obtain a pure quadratic form

$$(e_0 - \mu)b_0^+ b_0 + \frac{1}{2}c(b_0^+ b_0^+ + b_0 b_0).$$

The vanishing of the linear term imposes the constraint

$$e_0 - \mu - w\sqrt{V}n_0^{\frac{1}{3}} = 0. \tag{7.10}$$

The diagonalization of the bi-linear terms is achieved again by a canonical transformation and the excitation energy of the newly defined boson of $k = 0$ is vanishing due to the new s.c. constraint 7.10

$$\epsilon_0 = \sqrt{(e_0 - \mu)^2 - |c|^2} = 0$$

Therefore finally we get the ground state excitation spectrum for any k as

$$\epsilon_k = \sqrt{(\frac{\hbar^2 k^2}{2m} + w n_0)^2 - (w n_0)^2}. \tag{7.11}$$

This spectrum behaves linearly for $k \to 0$ as $\epsilon_k \approx \hbar\sqrt{\frac{w n_0}{m}}k$ and behaves quadratic like the free bosons for $k \to \infty$.

The phonon-like behavior as $k \to 0$ is in agreement with the Goldstone theorem stating that in the case of symmetry breaking of the ground state (here the simple phase transformation implying particle number conservation) the bosonic excitation spectrum must vanish at $k = 0$. One must stress, however, that the generalization of this theorem by Hugenholtz and Pines for finite temperatures cannot be fulfilled by the corresponding generalization of the Bogoliubov model.

7.4 Time-Dependent Bogoliubov and Gross–Pitaevskii Equations

The self-consistent Bogoliubov model may be extended also to a time-dependent version. One keeps the Hamiltonian equation 7.4, but the self-consistent averages are considered time-dependent according to the Hartree–Fock Schrödinger equation with some initial condition. Therefore, due to the self-consistency requirement, the Hamiltonian itself is time dependent.

$$
\begin{aligned}
H_B(t) = \int d\mathbf{x} \Bigg\{ & \psi(\mathbf{x})^+ \left(-\frac{\hbar^2}{2m}\nabla^2 + \right) \psi(\mathbf{x}) \\
& + \frac{w}{2} \Big[4\psi(\mathbf{x})^+\psi(\mathbf{x})\langle\psi(\mathbf{x})^+\psi(\mathbf{x})\rangle_t \\
& + \left(\psi^+(\mathbf{x})\psi^+(\mathbf{x})\langle\psi(\mathbf{x})\rangle_t^2 + h.c. \right) \\
& - \left(4\psi(\mathbf{x})\langle\psi(\mathbf{x})^+\rangle_t |\langle\psi(\mathbf{x})\rangle_t|^2 + h.c. \right) \Big] \Bigg\} + C_B(t).
\end{aligned}
\tag{7.12}
$$

The Schrödinger equation for the second quantized wave function $\psi(\mathbf{x}, t)$ is

$$
\begin{aligned}
\imath\, \hbar \frac{\partial}{\partial t}\psi(\mathbf{x}, t) = & -\frac{\hbar^2}{2m}\nabla^2\psi(\mathbf{x}, t) \\
+ \frac{w}{2}\Big[& 4\langle\psi(\mathbf{x}, t)^+\psi(\mathbf{x}, t)\rangle\psi(\mathbf{x}, t) + 2\langle\psi(\mathbf{x}, t)\rangle^2\psi^+(\mathbf{x}, t) - 4\langle\psi(\mathbf{x}, t)\rangle|\langle\psi(\mathbf{x}, t)\rangle|^2 \Big].
\end{aligned}
\tag{7.13}
$$

If we admit that $\langle\psi\psi\rangle = \langle\psi\rangle^2$ as it was admitted in the derivation of the Bogoliubov model, then the average particle number $\langle N\rangle$ is conserved. The energy conservation may be always fitted with $C_B(t)$ under the same assumption.

The equation for the condensate wave function $\langle\psi(\mathbf{x}, t)\rangle$

$$
\begin{aligned}
\imath\hbar\frac{\partial}{\partial t}\langle\psi(\mathbf{x}, t)\rangle = & -\frac{\hbar^2}{2m}\nabla^2\langle\psi(\mathbf{x}, t)\rangle \\
& + w\left[2\langle\psi(\mathbf{x}, t)^+\psi(\mathbf{x}, t)\rangle\langle\psi(\mathbf{x}, t)\rangle - \langle\psi(\mathbf{x}, t)\rangle|\langle\psi(\mathbf{x}, t)\rangle|^2 \right]
\end{aligned}
\tag{7.14}
$$

follows. It is coupled to the average density $\langle\psi(\mathbf{x}, t)^+\psi(\mathbf{x}, t)\rangle$ and leads to a strange mathematical problem. The equation for this average by the continuity equation needs also the knowledge of the average $\langle\psi(\mathbf{x}, t)^+\psi(\mathbf{x}', t)\rangle$, also of a two-point function with a singular initial condition $\langle\psi(\mathbf{x}, 0)^+\psi(\mathbf{x}', 0)\rangle = \delta(\mathbf{x} - \mathbf{x}')\langle\psi(\mathbf{x}, 0)^+\psi(\mathbf{x}, 0)\rangle$.

However, if one makes one more approximation, namely

$$
\langle\psi(\mathbf{x}, t)^+\psi(\mathbf{x}, t)\rangle \approx |\langle\psi(\mathbf{x})\rangle|^2,
$$

i.e., that the whole density consist only of condensate (as we considered also in the previous Subsection), then one gets the closed Gross–Pitaevskii equation for the condensate

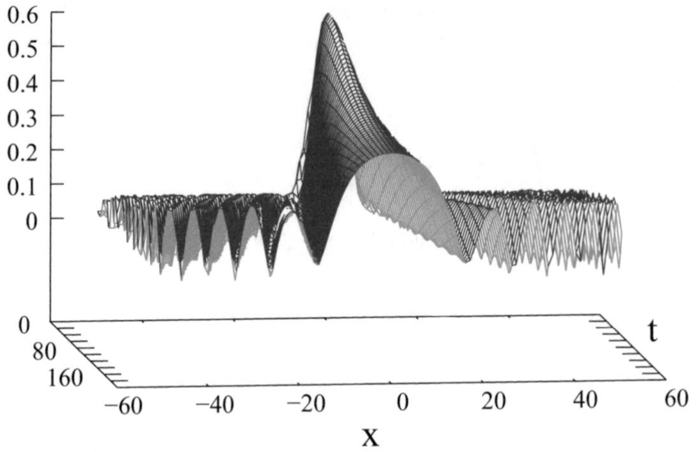

Fig. 7.4 The solution $|\langle\psi(x,t)\rangle|$ of the time-dependent 1D Gross–Pitaevskii equation with an initial Gaussian peak

$$\imath\hbar\frac{\partial}{\partial t}\langle\psi(\mathbf{x},t)\rangle = -\frac{\hbar^2}{2m}\nabla^2\langle\psi(\mathbf{x},t)\rangle + w|\langle\psi(\mathbf{x})\rangle|^2\langle\psi(\mathbf{x},t)\rangle. \tag{7.15}$$

This non-linear Schrödinger equation conserves the total number of condensate particles $\int d\mathbf{x}|\langle\psi(\mathbf{x})\rangle|^2$. Due to its non-linearity it gives rise to structure building typical for non-linear phenomena.

We illustrate this feature on the example of the solution of the Gross–Pitaevskii equation in 1D with a Gaussian initial condition for $|\langle\psi(\mathbf{x},t)\rangle|$ as it develops a soliton-like behavior in time. One sees in Fig. 7.4 the decrease of the original peak and the rise of satellite peaks in the time evolution.

7.5 Superconductivity

A phase transition of special kind is superconductivity of some materials below a certain critical temperature. Its peculiarity is that it reveals not only spectacular equilibrium properties like the Meissner effect (ideal diamagnetism expelling magnetic fields), but also non-usual non-equilibrium properties as the flow of electric current without resistance. Besides, connectivity properties of the material by allowing circulating permanent currents also play a decisive role. We shall describe in what follows the phenomenological theory of London which succeeds in a simple mathematical formulation to reflect the essentials of the phenomena. Thereafter we discuss a quantum mechanical model in real space, suggested by the BCS (Bardeen, Cooper, Schrieffer) theory of an effective electron–electron attraction mediated by phonons.

7.5.1 The Phenomenological Theory of London

The macroscopic Maxwell equations for the electric and magnetic fields and their sources (charge and current densities) are

$$\begin{aligned}
\nabla\mathbf{B} &= 0\\
\nabla\times\mathbf{E} &= -\frac{1}{c}\frac{\partial}{\partial t}\mathbf{B}\\
\nabla\times\mathbf{B} &= \frac{4\pi}{c}\mathbf{j}+\frac{1}{c}\frac{\partial}{\partial t}\mathbf{E}\\
\nabla\mathbf{E} &= 4\pi\rho.
\end{aligned}$$

The charges and currents obviously have to satisfy the conservation rule

$$\nabla\mathbf{j}+\frac{\partial}{\partial t}\rho=0.$$

These equations alone do not determine the macroscopic electromagnetic fields. They have to be supplemented by "matter relations" for the current- and charge-densities characteristic for the given material, which relates these entities again to the fields.
The simplest examples are:

- *normal conductors* (metals), with no charge density inside and Ohms law for the current density

$$\mathbf{j}_{Ohm}=\sigma\mathbf{E},\qquad \rho=0.$$

- *insulators (dielectrics)*, with polarization charges and polarization currents characterized by a dielectric constant

$$\mathbf{j}_{pol}=\frac{1}{4\pi}(\epsilon-1)\frac{\partial}{\partial t}\mathbf{E},\qquad \rho=-\frac{1}{4\pi}(\epsilon-1)\nabla\mathbf{E}.$$

- *semiconductors*, which beside the ohmic and polarization currents also have a diffusion current induced by the "free" charge density ρ_{free}

$$\mathbf{j}_{diff}=-D\nabla\rho_{free}.$$

  The diffusion constant D is related to the conductivity by the Einstein relation defined by the carrier density n and the temperature

$$\sigma=ne^2k_BTD.$$

– *magnetic materials* with the specific magnetization current density

$$\mathbf{j}_{magn} = c\nabla \times \mathbf{M},$$

where the relationship $\mathbf{M}(\mathbf{B})$ may be non-linear (ferromagnets).

To describe the basic aspects of a superconductor, London proposed two simple phenomenological "matter relations" that indeed successfully describe the main properties of these materials.

The stipulated equations that relate the current density to the electromagnetic field are characterized by a phenomenological constant Λ

$$\mathbf{E} = \Lambda \frac{\partial}{\partial t}\mathbf{j} \tag{7.16}$$

$$\Lambda \nabla \times \mathbf{j} = -\frac{1}{c}\mathbf{B}. \tag{7.17}$$

The first London equation was suggested by the behavior of free electrons that are accelerated by an electric field. If one takes the curl of this equation and uses the Maxwell equations, one gets

$$\frac{\partial}{\partial t}\left(\Lambda \nabla \times \mathbf{j} + \frac{1}{c}\mathbf{B}\right) = 0.$$

Therefore, the second London equation states only that not just the time derivative of the expression under the bracket is vanishing, but the expression itself.
Now, taking the curl of the third Maxwell equation

$$\nabla \times \nabla \times \mathbf{B} = \frac{4\pi}{c}\nabla \times \mathbf{j} + \frac{1}{c}\frac{\partial}{\partial t}\nabla \times \mathbf{E}$$

and using this second London equation, as well as the identity

$$\nabla \times \nabla \times \mathbf{B} = \nabla(\nabla\mathbf{B}) - \nabla^2\mathbf{B}$$

one gets

$$\nabla^2\mathbf{B} - \frac{4\pi}{c^2\Lambda}\mathbf{B} - \frac{1}{c^2}\frac{\partial^2}{\partial t^2}\mathbf{B} = 0.$$

In a stationary regime with constant fields it looks as

$$\nabla^2\mathbf{B} - \frac{4\pi}{c^2\Lambda}\mathbf{B} = 0. \tag{7.18}$$

The length $\ell = c\sqrt{\frac{\Lambda}{4\pi}}$ obviously defines the penetration depth of a magnetic field inside a superconductor in equilibrium (Meissner effect). (Compared with the penetration of a static electric field into a semiconductor.)

After integrating the second London equation on a surface S

$$\int_S d\mathbf{s}\nabla \times \mathbf{j} = -\frac{1}{\Lambda c}\int_S d\mathbf{s}\mathbf{B}$$

and using (Stokes theorem) we get

$$\oint d\mathbf{l}\ \mathbf{j} = -\frac{1}{\Lambda c}\Phi \tag{7.19}$$

also, a circulation of the current around a loop may be sustained by a magnetic flux inside the loop. Since in equilibrium there is no magnetic field inside the superconductor only the multiple-connectivity of the sample allows for permanent circular currents.

These two main properties characterize the equilibrium properties of a superconductor. The first London equation describes non-equilibrium phenomena. These are much difficult to treat. The microscopical theories of superconductivity concentrate mostly on the equilibrium properties: the existence of a phase transition and the Meissner effect.

7.6 Superconducting Phase Transition in a Simple Model of Electron–Electron Interaction

The fundamental idea for a microscopical explanation of superconductivity stems from Bardeen–Cooper–Schrieffer (BCS) about the effective interaction between the electrons of opposite spins that might become attractive, at least in a limited range of momenta (in the neighborhood of the Fermi energy), due to exchange of optical phonons. This attraction produces no bound states, but just a correlation between electrons of opposite spins and momenta, with anomalous electron number non-conserving averages analogously to the former description of the Bose condensation. This BCS theory offers a correct interpretation of the properties of the superconducting phase transition.

On the other hand, the theory of the electromagnetic properties must be formulated in the real space. Here we want to explore the implementation of the BCS idea in the Rickayzen model based on an electron–electron interaction by a fictitious potential without any claim for a realistic theory of superconductivity. We shall treat within this approach also the Meissner effect.

We consider instead of the Hamilton operator H, the operator $\mathcal{H} \equiv H - \mu N$

$$\begin{aligned}\mathscr{H} &\equiv H - \mu N\\ &= \sum_{\sigma=\pm\frac{1}{2}} \int d\mathbf{x}\psi_\sigma^+(\mathbf{x}) \left\{ \frac{1}{2m}\left(-\imath\hbar\nabla - \frac{e}{c}\mathbf{A}(\mathbf{x})\right)^2 - \mu \right\} \psi_\sigma(\mathbf{x})\\ &+ \frac{1}{2}\int d\mathbf{x} \int d\mathbf{x}' V(\mathbf{x}-\mathbf{x}')\{\psi^+_{\frac{1}{2}}(\mathbf{x})\psi^+_{-\frac{1}{2}}(\mathbf{x}')\psi_{\frac{1}{2}}(\mathbf{x}')\psi_{-\frac{1}{2}}(\mathbf{x}) + h.c.\}\end{aligned}$$

with some "attractive" potential $V(\mathbf{x})$ in the presence of a constant magnetic field described by the vector potential $\mathbf{A}(\mathbf{x})$ in the Coulomb gauge $\nabla\mathbf{A}(\mathbf{x}) = 0$. The chosen interaction does not lead to bound states, like an attractive Coulomb potential would do, but, as we shall see, it gives rise to BCS-type correlations.

We approximate the problem within the Hartree–Fock–Bogoliubov scheme by retaining only the anomalous (electron number non-conserving) averages with opposite spins

$$\begin{aligned}\mathscr{H}_{HFB} &= \sum_{\sigma} \int d\mathbf{x}\psi_\sigma^+(\mathbf{x}) \left\{ \frac{1}{2m}\left(-\imath\hbar\nabla - \frac{e}{c}\mathbf{A}(\mathbf{x})\right)^2 - \mu \right\} \psi_\sigma(\mathbf{x})\\ &+ \frac{1}{2}\int d\mathbf{x} \int d\mathbf{x}' V(\mathbf{x}-\mathbf{x}') \Big\{ \langle\psi^+_{\frac{1}{2}}(\mathbf{x})\psi^+_{-\frac{1}{2}}(\mathbf{x}')\rangle\psi_{-\frac{1}{2}}(\mathbf{x}')\psi_{\frac{1}{2}}(\mathbf{x})\\ &+ \langle\psi_{-\frac{1}{2}}(\mathbf{x}')\psi_{\frac{1}{2}}(\mathbf{x})\rangle\psi^+_{\frac{1}{2}}(\mathbf{x})\psi^+_{-\frac{1}{2}}(\mathbf{x}') - \langle\psi^+_{\frac{1}{2}}(\mathbf{x})\psi^+_{-\frac{1}{2}}(\mathbf{x}')\rangle\langle\psi_{-\frac{1}{2}}(\mathbf{x}')\psi_{\frac{1}{2}}(\mathbf{x})\rangle \Big\}.\end{aligned}$$

Here the average is defined over the macro-canonical distribution

$$\langle\ldots\rangle \equiv \frac{Tr\left\{e^{-\beta\mathscr{H}_{HFB}}\ldots\right\}}{Tr\left\{e^{-\beta\mathscr{H}_{HFB}}\right\}}.$$

In the absence of the magnetic field ($\mathbf{A} = 0$), in a finite volume Ω with discrete wave vectors we get the unperturbed "Hamiltonian" in terms of the creation and annihilation operators of the $\mathbf{k}, \sigma$ states

$$\begin{aligned}\mathscr{H}^0{}_{HFB} = \sum_{\mathbf{k}} \Big\{ &\epsilon_{\mathbf{k}} \left(a^+_{\mathbf{k},\frac{1}{2}} a_{\mathbf{k},\frac{1}{2}} + a^+_{-\mathbf{k},-\frac{1}{2}} a_{-\mathbf{k},-\frac{1}{2}} \right)\\ &+ \Delta_{\mathbf{k}} a_{-\mathbf{k},-\frac{1}{2}} a_{\mathbf{k},\frac{1}{2}} + \Delta^*_{\mathbf{k}} a^+_{\mathbf{k},\frac{1}{2}} a^+_{-\mathbf{k},-\frac{1}{2}} + \mathscr{C}(\mathbf{k}) \Big\}\end{aligned}$$

with

$$\epsilon(\mathbf{k}) = \frac{\hbar^2 k^2}{2m} - \mu,$$

$$\Delta_{\mathbf{k}} \equiv \frac{1}{\Omega} \sum_{\mathbf{k}'} V(\mathbf{k} - \mathbf{k}') \langle a^{+}_{\mathbf{k}', \frac{1}{2}} a^{+}_{-\mathbf{k}', -\frac{1}{2}} \rangle$$

and

$$\tilde{V}(\mathbf{k}) = \int_{\Omega} d\mathbf{x} e^{\imath \mathbf{k}\mathbf{x}} V(\mathbf{x}).$$

This “Hamiltonian” may be diagonalized with the Bogoliubov–Valatin canonical transformation

$$\begin{pmatrix} a_{\mathbf{k}, \frac{1}{2}} \\ a^{+}_{-\mathbf{k}, -\frac{1}{2}} \end{pmatrix} = \begin{pmatrix} u_{\mathbf{k}} & v_{\mathbf{k}} \\ -v^{*}_{\mathbf{k}} & u^{*}_{\mathbf{k}} \end{pmatrix} \begin{pmatrix} c_{\mathbf{k}, \frac{1}{2}} \\ c^{+}_{-\mathbf{k}, -\frac{1}{2}} \end{pmatrix},$$

where the coefficients must satisfy the relation

$$|u_k|^2 + |v_k|^2 = 1.$$

In terms of the new creation–annihilation operators, we get

$$\mathscr{H}^{0}_{HFB} = \sum_{\mathbf{k}} \left\{ E_{\mathbf{k}} \left(c^{+}_{\mathbf{k}, \frac{1}{2}} c_{\mathbf{k}, \frac{1}{2}} + c^{+}_{\mathbf{k}, -\frac{1}{2}} c_{\mathbf{k}, -\frac{1}{2}} \right) + \mathscr{C}(\mathbf{k}) \right\},$$

where

$$E(\mathbf{k}) = \sqrt{\epsilon(\mathbf{k})^2 + |\Delta(\mathbf{k})|^2},$$

and (in the infinite volume limit)

$$\Delta(\mathbf{k}) = -\frac{1}{(2\pi)^3} \int d\mathbf{k}' \tilde{V}(\mathbf{k} - \mathbf{k}') \frac{\Delta(\mathbf{k}')}{2E(\mathbf{k}')} \left[1 - 2f\left(E(\mathbf{k}')\right)\right]$$

with the Fermi function

$$f(E) = \frac{1}{e^{\beta E} + 1}.$$

The HFB constant is

$$\mathscr{C}(\mathbf{k}) = \epsilon_{\mathbf{k}} - E_{\mathbf{k}} + \frac{|\Delta_{\mathbf{k}}|^2}{2E_{\mathbf{k}}} \left[1 - 2f\left(E(\mathbf{k})\right)\right].$$

The chemical potential μ has to be determined from the equation for the average density of electrons that in the new variables looks as

$$\frac{1}{(2\pi)^3}\int d\mathbf{k}\left[1-\frac{\epsilon(\mathbf{k})}{E(\mathbf{k})}[1-2f(E(\mathbf{k}))]\right]=\langle n\rangle.$$

The grand-canonical potential is

$$\mathscr{F}=-\frac{1}{\Omega\beta}\ln\left\{Tr\left[e^{-\beta\mathscr{H}^0_{HFB}}\right]\right\}=\frac{1}{(2\pi)^3}\int d\mathbf{k}\left\{\frac{2}{\beta}\ln[1-f(E(\mathbf{k})]+\mathscr{C}(\mathbf{k})\right\}$$

and the solution of the gap equation ensures also the extremum condition

$$\frac{\delta\mathscr{F}}{\delta\Delta(\mathbf{k})}=0,$$

while the stability of the phase transition depends on the sign of its second derivative.

The gap equation obviously has a vanishing solution $\Delta(\mathbf{k}) = 0$ and possibly another one with $\Delta(\mathbf{k}) \neq 0$. If the solution with $\Delta(\mathbf{k}) \neq 0$ gives rise to a smaller grand-canonical potential, the anomalous solution is preferred, and we have a phase transition. In this case the new ground state is the new vacuum, with energy $\mathscr{C}(\mathbf{k})$, while the next excited state with a quasi-particle of wave vector $\mathbf{k}$ has the energy $\mathscr{C}(\mathbf{k}) + E(\mathbf{k})$. The minimal distance between the two states may be finite and we have an energy gap for excitations. (Due to the rotational invariance of the potential V, not only the unperturbed energies ϵ but also the new ones E as well as Δ are functions only of $k \equiv |\mathbf{k}|$ and we may omit the vector notation.)

We have avoided to define the potential $V(\mathbf{x})$ itself and it is not at all clear if the BCS theory may be formulated explicitly as an attraction through a local potential in real space. Therefore, the above discussion serves only as a sketch of a possible scenario for a phase transition with a BCS-type electron correlation. For any given potential one may check (numerically) the expected properties.

Whether the above described theory or any similar one describes indeed a superconductor may be checked only by understanding its electromagnetic properties. A theory of the resistanceless flow of currents is beyond any hope. Indeed, today's transport theory introduces elements of irreversibility by hand in order to get a finite conductivity. If one omits this step, then any model allows free flow of electric current. Remains the question, what would impede dissipation? One may hope that the gap above the Fermi energy offers stability against weak perturbations. (However, we have seen on the example of a solvable model in Sect. 5.2 that indeed even in an open system dissipation may be absent in a certain parameter range.) The only well defined possibility remains to look at least at magnetic properties in equilibrium.

7.6.1 Meissner Effect Within Equilibrium Linear Response

The simplest test of superconducting features in the above described model is to see whether it gives rise to the Meissner effect, i.e., to expulsion of magnetic fields from the bulk? Since the Meissner effect occurs in equilibrium, one may use the equilibrium linear response theory.

The basic Hamiltonian used in solid state theory omits terms of order $\frac{1}{c^2}$ and this means not only omitting relativistic corrections, but also ignoring velocity–velocity couplings between the electrons dictated by electrodynamics. Thus, the magnetic field produced by the electrons is completely absent. Therefore, we are compelled to interpret here the magnetic field as an s.c. (mean-field) one, in analogy to the treatment of longitudinal electric fields in a system of electrons without explicit Coulomb interaction.

Let us consider a weak magnetic field that may be considered as a small perturbation. The peculiarity of the self-consistent approach here is that the vector potential $\mathbf{A}(\mathbf{x})$ contained explicitly in the kinetic energy is not the only perturbation due to the magnetic field. The anomalous averages in the s.c. "Hamiltonian" $\mathscr{H}_{HFB}$ are also modified by the magnetic field. Therefore, the perturbation to consider for the linear response is

$$\mathscr{H}' = -\frac{1}{c}\int d\mathbf{x}\mathbf{j}(\mathbf{x})\mathbf{A}(\mathbf{x}) + \int d\mathbf{x}\int d\mathbf{x}' V(\mathbf{x}-\mathbf{x}')\left[\eta(\mathbf{x},\mathbf{x}')\psi_{-\frac{1}{2}}(\mathbf{x}')\psi_{\frac{1}{2}}(\mathbf{x}) + h.c.\right],$$

where

$$\eta(\mathbf{x},\mathbf{x}') \equiv \langle\psi^+_{\frac{1}{2}}(\mathbf{x})\psi^+_{-\frac{1}{2}}(\mathbf{x}')\rangle - \langle\psi^+_{\frac{1}{2}}(\mathbf{x})\psi^+_{-\frac{1}{2}}(\mathbf{x}')\rangle_0$$

is (the first order in the magnetic field) deviation of the averages from their unperturbed values. These must be simultaneously calculated self-consistently. In the presence of a magnetic field the current density contains also a diamagnetic term

$$\mathbf{j}(\mathbf{x}) \equiv \frac{e}{2m}\psi^+(\mathbf{x})\left(-\imath\hbar\nabla + \frac{e}{c}\mathbf{A}(\mathbf{x})\right)\psi(\mathbf{x}) + h.c.,$$

(however, see Chap. 9 for the proper formulation, taking into account the difference between the s.c. mean and the external fields).

One may use the equilibrium linear response formula Eq. 6.1 with the previous results for the chosen model of attracting electrons to get the linear relationship between the average current and the vector potential.

From the current conservation it follows that

$$\langle\tilde{j}_\mu(\mathbf{k})\rangle = \left(\delta_{\mu\nu} - \frac{k_\mu k_\nu}{k}\right)\kappa(k)\tilde{A}_\nu(\mathbf{k}).$$

Due to the transversality of the vector potential in the Coulomb gauge this is equivalent to

$$\langle \tilde{j}_\mu(\mathbf{k})\rangle = \kappa(k)\tilde{A}_\mu(\mathbf{k}).$$

The coefficient $\kappa(k)$ given by the equilibrium linear response theory has a rather complicated expression in terms of the previously defined parameters and we omit it. Actually, we are interested in this relationship only for small wave vectors (slowly varying behavior in the coordinate space!). In the anomalous superconducting phase having a non-vanishing gap $\Delta(\mathbf{k})$ at $\mathbf{k} = 0$ one gets

$$\kappa(0) = -\frac{1}{c\Lambda}.$$

It may be shown that under the condition of the stability of the superconducting phase Λ is indeed positive and the contribution produced by consideration of $\eta(\mathbf{x}, \mathbf{x}')$ to this result vanishes. Therefore, the above described model seems to produce a Meissner effect, since in the Coulomb gauge the second London equation reads as

$$\mathbf{j} = -\frac{1}{c\Lambda}\mathbf{A}.$$

An obvious weakness of the above discussion is that in the bulk the London equation is meaningless, since if a Meissner effect occurs, the magnetic field as well as the current density should vanish in the bulk.

However, another word of caution is compulsory. The Meissner effect obviously is due to the compensation of the external magnetic field by the internal one. However, in this mean-field interpretation of the magnetic field in bulk one cannot differentiate these two. Here we are confronted with the limits of today's solid state theory based on a pure Coulomb interacting model of electrons and ions. See, however, the discussion in Chap. 9 for a correct formulation of the Meissner effect in the bulk. The best way, however, is still to formulate the problem at the surface.

7.6.2 The Case of a Contact Potential: The Bogoliubov–de Gennes Equation

A simplest case to discuss in more detail is the choice of a contact potential

$$V(\mathbf{x}) = -g\delta(\mathbf{x}). \tag{7.20}$$

This implies a $\mathbf{k}$-independent gap satisfying the equation

$$\Delta = \frac{g}{(2\pi)^3} \int d\mathbf{k} \frac{\Delta}{2E(\mathbf{k})} [1 - 2f(E(\mathbf{k}))]$$

which besides Δ=0 has no other solution, except if one cuts off the integral at some k_M. This cutoff has then to be taken into account in the definitions of all the summations over the **k**-space and the choice $\tilde{V}(\mathbf{k}) = -g\theta(k_M - k)$ implies

$$V(\mathbf{x}) = \frac{1}{(2\pi)^3} \int d\mathbf{k}\theta(k_M - k)e^{\imath \mathbf{k}\mathbf{x}} = \frac{\sin(k_M r) - k_M r \cos(k_M r)}{2\pi^2 r^3} \quad (7.21)$$

which is an approximant of the $\delta(\mathbf{x})$ function and not a true contact potential as it was assumed in Eq. 7.20.

Another approach, due to de Gennes is to remain in the case of the contact potential within the coordinate space formulation. The Hartree–Fock–Bogoliubov–de Gennes Hamiltonian with the anomalous averages in the coordinate space in the presence of a vector potential $\mathbf{A}(\mathbf{x})$ and a scalar potential $U(\mathbf{x})$ looks as

$$\begin{aligned}\mathscr{H}_{Genn} = &\sum_\sigma \int d\mathbf{x}\psi_\sigma^+(\mathbf{x}) \left\{ \frac{1}{2m}\left(-\imath\hbar\nabla - \frac{e}{c}\mathbf{A}(\mathbf{x})\right)^2 + U(\mathbf{x}) - \mu \right\} \psi_\sigma(\mathbf{x}) \\ &- \frac{g}{2}\int d\mathbf{x}\Big\{ \langle \psi_{\frac{1}{2}}^+(\mathbf{x})\psi_{-\frac{1}{2}}^+(\mathbf{x})\rangle \psi_{-\frac{1}{2}}(\mathbf{x})\psi_{\frac{1}{2}}(\mathbf{x}) + \langle \psi_{-\frac{1}{2}}(\mathbf{x})\psi_{\frac{1}{2}}(\mathbf{x})\rangle \psi_{\frac{1}{2}}^+(\mathbf{x})\psi_{-\frac{1}{2}}^+(\mathbf{x}) \\ &- \langle \psi_{\frac{1}{2}}^+(\mathbf{x})\psi_{-\frac{1}{2}}^+(\mathbf{x})\rangle\langle \psi_{-\frac{1}{2}}(\mathbf{x})\psi_{\frac{1}{2}}(\mathbf{x})\rangle \Big\}\end{aligned} \quad (7.22)$$

or with the notation

$$\mathscr{D}(\mathbf{x}) = \langle \psi_{-\frac{1}{2}}(\mathbf{x})\psi_{\frac{1}{2}}(\mathbf{x})\rangle, \quad (7.23)$$

$$\begin{aligned}\mathscr{H}_{Genn} = &\sum_\sigma \int d\mathbf{x}\psi_\sigma^+(\mathbf{x}) \left\{ \frac{1}{2m}\left(-\imath\hbar\nabla - \frac{e}{c}\mathbf{A}(\mathbf{x})\right)^2 + U(\mathbf{x}) - \mu \right\} \psi_\sigma(\mathbf{x}) \\ &- \frac{g}{2}\int d\mathbf{x}\Big\{ \mathscr{D}(\mathbf{x})^* \psi_{-\frac{1}{2}}(\mathbf{x})\psi_{\frac{1}{2}}(\mathbf{x}) + \mathscr{D}(\mathbf{x})\psi_{\frac{1}{2}}^+(\mathbf{x})\psi_{-\frac{1}{2}}^+(\mathbf{x}) + |\mathscr{D}(\mathbf{x})|^2 \Big\}.\end{aligned} \quad (7.24)$$

It may be diagonalized by a Bogoliubov transformation to new fermion operators

$$[\alpha_{n\sigma}, \alpha_{n',\sigma'}^+] = \delta_{nn'}\delta_{\sigma,\sigma'}$$

by writing

$$\psi_{\frac{1}{2}}(\mathbf{x}) = \sum_n \left\{ u_n(\mathbf{x})\alpha_{n,\frac{1}{2}} + v_n(\mathbf{x})^* \alpha_{n,-\frac{1}{2}}^+ \right\}$$

$$\psi_{-\frac{1}{2}}(\mathbf{x}) = \sum_n \left\{ u_n(\mathbf{x})\alpha_{n,-\frac{1}{2}} - v_n(\mathbf{x})^*\alpha^+_{n,\frac{1}{2}} \right\}.$$

Thus

$$\mathscr{H}_{Genn} = E_0 + \sum_{n,\sigma} E_n \alpha^+_{n,\sigma}\alpha_{n,\sigma}$$

while the eigenenergies E_n satisfy the coupled eigenvalue equations

$$Eu_n(\mathbf{x}) = \left[\frac{\hbar^2}{2m}\left(-\imath\hbar\nabla - \frac{e}{c}\mathbf{A}(\mathbf{x})\right)^2 + U(\mathbf{x}) - \mu\right]u_n(\mathbf{x}) + g\mathscr{D}(\mathbf{x})v_n(\mathbf{x}) \quad (7.25)$$

$$Ev_n(\mathbf{x}) = -\left[\frac{\hbar^2}{2m}\left(-\imath\hbar\nabla - \frac{e}{c}\mathbf{A}(\mathbf{x})\right)^2 + U(\mathbf{x}) - \mu\right]v_n(\mathbf{x}) + g\mathscr{D}(\mathbf{x})^*u_n(\mathbf{x}) \quad (7.26)$$

and the self-consistency condition

$$\mathscr{D}(\mathbf{x}) = g\sum_n v_n(\mathbf{x})^*u_n(\mathbf{x})\tanh(\beta E_n/2), \quad (7.27)$$

which for $T = 0$ looks simpler

$$\mathscr{D}(\mathbf{x}) = g\sum_n v_n(\mathbf{x})^*u_n(\mathbf{x})\theta(E_n).$$

Again it is necessary to limit the energy spectrum from above ($E < E_M$), but one got a description in real space that allows formulation of boundary conditions! If one admits that the system is a half space bounded by a plane at $z = 0$, i.e., $U(x, y, 0) = \infty$ and vanishing otherwise, then due to the vanishing boundary conditions $\mathscr{D}(0, y, z) = 0$.

Obviously the de Genes theory including boundary conditions differs from the bulk theory with the pseudo-contact potential Eq. 7.21.

On the other hand, the vector potential considered by de Gennes should be considered again as the self-consistent one inside the superconductor. This is the sole one modified within the material, not the applied one. (As we shall later discuss it in Chap. 9 only the linear version in the s.c. field of the de Gennes Hamiltonian is correct from the point of view of quantum electrodynamics.) A proof of the Meissner effect within the s.c. frame has to consider the electromagnetic boundary problem with an external (applied) magnetic field having its sources far away from the system. The formulation of de Gennes offers this possibility. The required ingredient has to be a quasi-local relation between the average current density and the s.c. vector potential in the material. Such a relation has to be valid already at distances smaller than the resulting penetration depth. Unfortunately such a relationship is hard to proof and may raise questions about possible long range correlations oft associated with phase transitions. Inclusion of other ingredients in the theory might, however, endanger the phase transition itself. The situation was quite different in the case of the Debye penetration of a static electric field into a semiconductor described in Sect. 2.6.1.

Chapter 8
Low Dimensional Semiconductors

Quasi two-dimensional semiconductor layers offer a lot of spectacular properties of which the most famous is the Quantum Hall effect. We limit ourselves, however, only to the presentation of the simplest, well understood, but nevertheless surprising peculiarities of the classical and quantum mechanical two-dimensional motion of Coulomb interacting electrons in the presence of a strong transverse magnetic field.

In the last decades progresses in semiconductor technology produced ultra-thin semiconductor layer systems in which by a suitable choice of the layers, the electrons and holes are restricted to a two-dimensional motion. An ultra-thin layer of a semiconductor is inserted between two thick layers of another semiconductor with a larger band gap and therefore, both the electrons and holes in the ultra-thin layer are in a quantum well and their lowest states are discrete as it is shown in Fig. 8.1.

At sufficiently low temperatures the electrons and holes sit on their lowest levels (shown here in red and green). It means, the transverse motion at low temperatures is "frozen" in its lowest lying state $\phi_0(z)$ in the potential well created by the adjoining layers and the wave function of an electron, in a good approximation is given by

$$\Psi(x, y, z) = \psi(x, y)\phi_0(z),$$

with the coordinates x, y lying in the plane and z being the transverse coordinate. This is the experimental realization of a two dimensional (2D) electron–hole system, which has a lot of very interesting properties.

Already the free motion in 2D has a peculiarity, namely the one-electron state density is constant above $\epsilon = 0$

L. A. Bányai, *A Compendium of Solid State Theory*,
https://doi.org/10.1007/978-3-030-37359-7_8

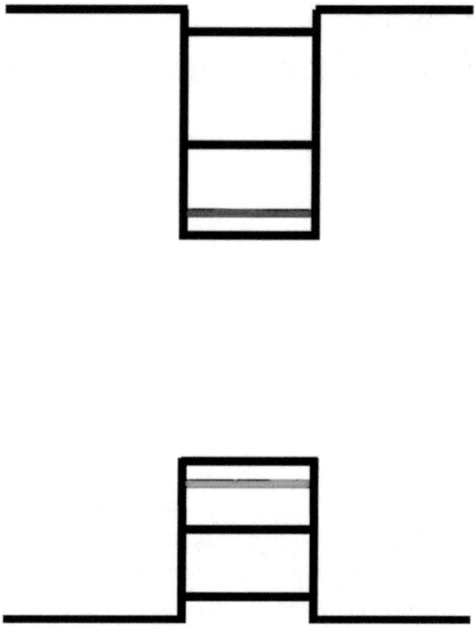

Fig. 8.1 A semiconductor quantum well

$$z(\epsilon) = \frac{1}{(2\pi)^2} \int d\mathbf{k}\delta(\epsilon - \frac{\hbar^2 k^2}{2m}) = \frac{m}{2\pi\hbar^2}\theta(\epsilon).$$

Such systems show above all spectacular properties in the presence of a strong magnetic field transverse to the plane. The most famous one is the Quantum Hall effect. We will not try to describe here the different, sometimes contradicting theories of this effect. Nevertheless, we want to bring the attention to the kind of strange physics we encounter in two dimensions (2D).

8.1 Exciton in 2D

Let us consider an electron–hole pair with Coulomb attraction in a 2D semiconductor. In the effective mass approximation, the relative motion in cylindrical coordinates is described by the Hamiltonian

$$H = -\frac{\hbar^2}{2m}\frac{1}{\rho}\frac{\partial}{\partial\rho}\left(\rho\frac{\partial}{\partial\rho}\right) - \frac{1}{\rho^2}\frac{\partial^2}{\partial\phi^2} + \frac{e^2}{\epsilon\rho},$$

where m is the reduced mass of the pair and ϵ is the dielectric constant (supposed here as being the same in the 2D layer as in the surrounding semiconductor). The eigenstates of this Hamiltonian are characterized by two quantum numbers $n =$

$0, 1, 2 \ldots$ and $\mu = \pm 0, \pm 1, \pm 2, \ldots$. The lowest eigenstate ($n = 0,\ \mu = 0$) is

$$\psi_{0,0} = \frac{4}{\sqrt{2\pi}a} e^{-\frac{2\rho}{a_B}},$$

while the ground state energy is

$$E_{0,0} = -4E_R.$$

where we used as parameter the Bohr radius a_B and the Rydberg energy E_R of the 3D exciton. A comparison with the ground state exciton wave function in 3D (see Sect. 3.1.2) shows that the exciton radius in 2D is twice smaller than in 3D and its binding energy is four times bigger. Contrary to any expectations, the transversely compressed exciton shrinks also in the still allowed two dimensions.

8.2 Motion of a 2D Electron in a Strong Magnetic Field

According to the discussed 3D motion of an electron in a homogeneous magnetic field (see Sect. 2.2.2) the stationary states in the Landau gauge are given by a plane wave in the field direction (along the z axis) and in the transverse plane by the Landau states of discrete energies. If one restricts the motion to the plane x,y, then the energy of such a Landau state is just

$$\epsilon_{n,X,} = \hbar\omega_c(n + \frac{1}{2}) \qquad (n = 0, 1, 2, \ldots).$$

Since these energies do not depend on the quantum number X (the x-coordinate of the center of the cyclotron motion), they are (in the absence of boundary conditions!) infinitely degenerate.

In what follows we consider very strong magnetic fields at very low temperatures and therefore we may consider that all the spins are aligned along the magnetic field. We shall consider the motion of such a 2D electron in a strong magnetic field perpendicular to the plane of motion in the presence of an external potential $U(\mathbf{r})$.

The simplest approach to consider is to ignore the higher lying Landau levels and, if the potential is weak, to consider its projection on the lowest lying Landau state ($n = 0$) within first order perturbation for the energy

$$E(X) = \frac{1}{2}\hbar\omega_0 + \langle 0, X|U(x, y)|0, X\rangle.$$

Kubo's more profound approach considers the limit of ultra-strong magnetic fields for arbitrary potentials and an arbitrary kinetic energy T. The 2D Hamilton operator is

$$H = T(\boldsymbol{\pi}) + U(\mathbf{r}),$$

where the generalized momenta $\boldsymbol{\pi}$ in the presence of the vector potential $\mathbf{A}$ are

$$\boldsymbol{\pi} \equiv \mathbf{p} + \frac{e}{c}\mathbf{A}(\mathbf{r})) \ ; \qquad \mathbf{p} \equiv \imath\hbar\nabla.$$

Within the set of operators $(\boldsymbol{\pi}, \mathbf{r})$ the only non-vanishing commutators are

$$[\pi_x, \pi_y] = \frac{\hbar e}{\imath c}B; \qquad [\pi_x, x] = [\pi_y, y] = \frac{\hbar}{\imath}.$$

Defining new operators

$$\xi = \frac{c}{eB}\pi_y, \quad \eta = \frac{c}{eB}\pi_x$$

and

$$X = x - \xi, \quad Y = y - \eta$$

it results that

$$[\xi, \eta] = \frac{\hbar c}{\imath eB} \equiv \frac{l_B^2}{\imath},$$

$$[X, Y] = -\frac{\hbar c}{\imath eB} \equiv -\frac{l_B^2}{\imath},$$

while the other commutators of the new operators vanish. The kinetic energy T depends only on the new operators ξ and η. Also, in the absence of the potential U the operators X and Y do not change in time. They are assimilated with the coordinates of the cyclotron motion, while the operators ξ and η are the relative coordinates.

In the presence of the external potential, either in the frame of quantum mechanics or within classical mechanics, the center of the cyclotron motion moves according to

$$\dot{X} = \frac{c}{eB}\frac{\partial U}{\partial y}, \qquad \dot{Y} = -\frac{c}{eB}\frac{\partial U}{\partial x}.$$

According to the above canonical commutation rules one has the uncertainty relations

$$\triangle X \triangle Y = 2\pi l_B^2$$

and since ultra-strong magnetic fields imply $l_B \to 0$, a classical description with classical coordinates X, Y is appropriate. On the other hand, if the potential is slowly varying on the scale of the magnetic length, one may ignore the relative motion and to a very good approximation one could use the classical equations of motion

$$\dot{X} = \frac{c}{eB}\frac{\partial U(X,Y)}{\partial Y}, \qquad \dot{Y} = -\frac{c}{eB}\frac{\partial U(X,Y)}{\partial X}.$$

It follows that

$$\frac{\partial Y}{\partial X} = \frac{\dot{Y}}{\dot{X}} = -\frac{\frac{\partial U(X,Y)}{\partial X}}{\frac{\partial U(X,Y)}{\partial Y}}.$$

However, this corresponds to the motion on a curve defined by $U(x, y) = U_0$. Indeed, by using the definition of the implicit derivative one gets

$$\frac{\partial y}{\partial x} = -\frac{\frac{\partial U(x,y)}{\partial x}}{\frac{\partial U(x,y)}{\partial y}}.$$

To conclude, in the limit of ultra-strong magnetic fields the motion of the cyclotron center of electrons in 2D with arbitrary kinetic energy in the presence of an external potential is just a classical one along the equipotential curves of the potential.

8.3 Coulomb Interaction in 2D in a Strong Magnetic Field

8.3.1 Classical Motion

A stranger aspect of the 2D motion in a strong magnetic field is how Coulomb forces act. Let us consider first the classical problem of the motion of two particles of opposite charges (electron and hole). We are interested only in the relative motion, therefore one of the particles we may keep fixed in the origin. A numerical solution of the corresponding Newton equations shows the trajectory in Fig. 8.2. This is a somewhat complicated picture, but the particles, as expected seem to attract each other and stick together.

To our surprise, even two identically charged particles in 2D, having repulsive Coulomb forces, stick together, showing an effective attraction, as it is illustrated in Fig. 8.3. Of course, the Coulomb force tries to accelerate the electrons in the repulsive manner, but the accelerated electron is returned by the bending in the magnetic field. The only escape would have been in the now forbidden transverse direction.

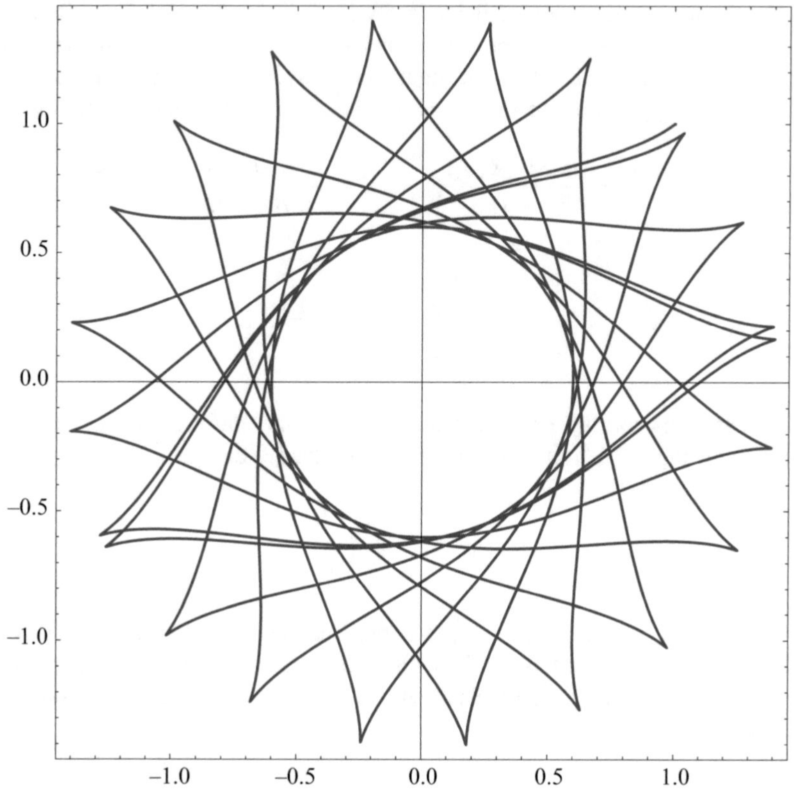

Fig. 8.2 Classical relative motion of opposite charged particles in 2D in the presence of a transverse magnetic field

If one considers a whole cluster of electrons these are sticking together in a cluster, as it is shown in Fig. 8.4.

8.3.2 Quantum Mechanical States

The quantum mechanical analysis of the motion of two Coulomb repulsive particles confirms also the existence of bound electron–electron states in 2D.
The quantum mechanical Hamiltonian for the 3D relative motion of two electrons in the presence of a magnetic field **B** is

$$H^{3D} = -\frac{\hbar^2}{2m}\nabla^2 - \frac{e\hbar}{mc}\imath\left(\mathbf{A}(\mathbf{r})\,.\,\nabla + \frac{1}{2}\nabla\mathbf{A}(\mathbf{r})\right) + \frac{e^2}{2mc^2}\mathbf{A}(\mathbf{r})^2 + \frac{e^2}{r}.$$

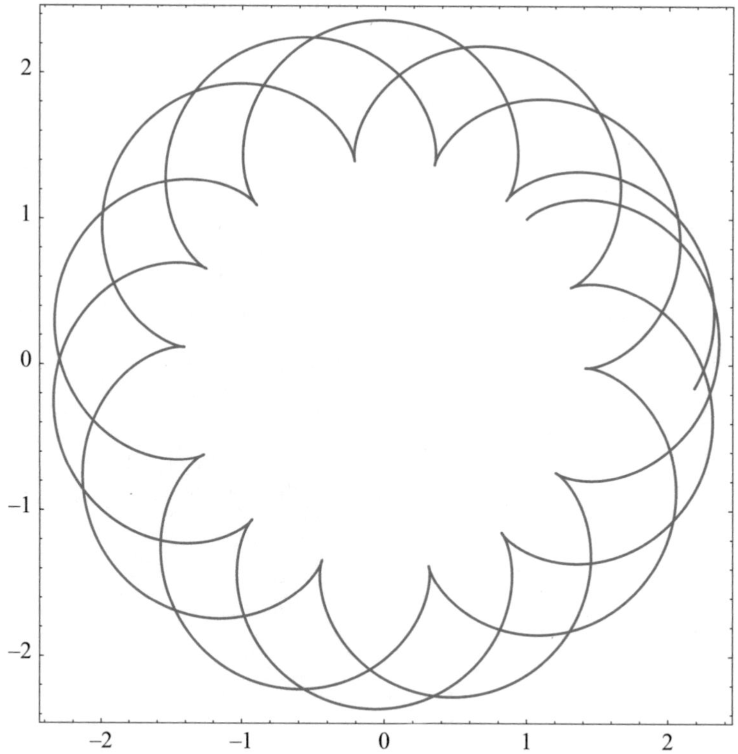

Fig. 8.3 Classical relative motion of identically charged particles in 2D electron in the presence of a transverse magnetic field

If one chooses the divergenceless vector potential (with the magnetic field in the z-direction)

$$\mathbf{A}(\mathbf{r}) \equiv \left(-\frac{1}{2}By,\ \frac{1}{2}Bx,\ 0\right),$$

one may write the 2D Hamiltonian describing the in-plane motion in cylinder coordinates $\rho,\ \phi$ as

$$H^{2D} = -\frac{\hbar^2}{2m}\frac{1}{\rho}\frac{\partial}{\partial\rho}\left(\rho\frac{\partial}{\partial\rho}\right) - \frac{1}{\rho^2}\frac{\partial^2}{\partial\phi^2} - \imath\frac{e\hbar B}{2mc}\frac{\partial}{\partial\phi} + \frac{e^2B^2}{8mc^2}\rho^2 + \frac{e^2}{\rho}.$$

One looks, as usual, for the eigenfunctions as

$$\psi(\rho, \phi) = u(\rho)e^{\imath\mu\phi} \qquad (\mu = 0, 1, 2, \ldots).$$

Then the eigenvalue problem for the radial part is

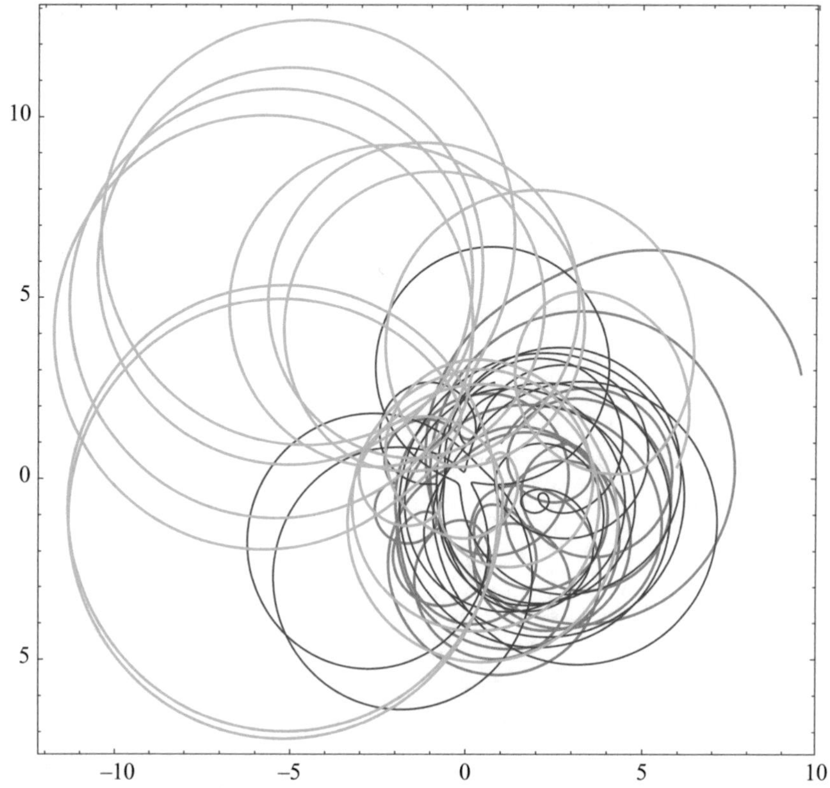

Fig. 8.4 Classical motion of 4 electrons (one is kept fixed in the origin) in 2D in the presence of the magnetic field

$$-\frac{\hbar^2}{2m}\frac{1}{\rho}\frac{\partial}{\partial\rho}\left(\rho\frac{\partial u}{\partial\rho}\right)+\left(\frac{\hbar^2}{2m}\frac{\mu^2}{\rho^2}+\frac{e\hbar B}{2mc}\mu+\frac{e^2B^2}{8mc^2}\rho^2+\frac{e^2}{\rho}\right)u=E_\mu u.$$

This Schrödinger equation differs (up to a shift in the energy with $\frac{e\hbar B}{2mc}\mu$) from that of a radial 2D oscillator just due to the Coulomb term $\frac{e^2}{\rho}$. Since in a very strong magnetic field the spins are supposed to be aligned along the magnetic field and the wave functions must be anti-symmetrical for fermions; only odd angular momenta μ are of interest. In terms of the dimensionless parameter

$$\xi=\frac{eB}{2c\hbar}\left(\frac{\hbar^2}{2me^2}\right)^2$$

an ultra-strong magnetic field corresponds to $\xi \gg 1$. The radial wave function of the lowest $\mu = 1$ state of the 2D oscillator is

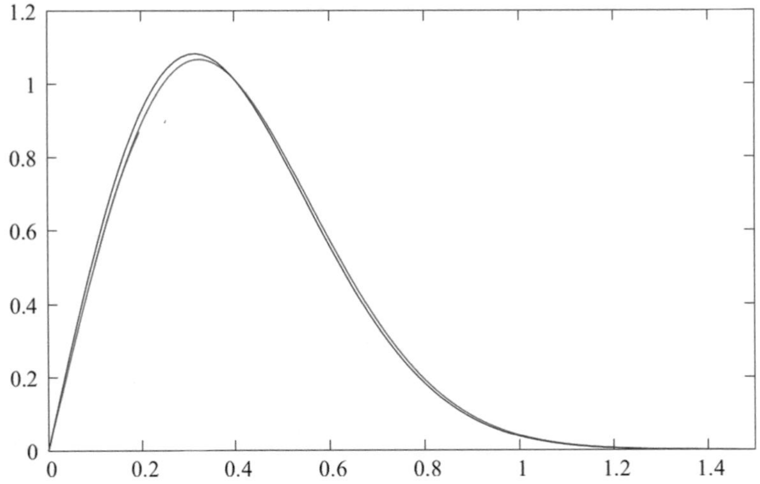

Fig. 8.5 The lowest $\mu = 1$ eigenfunction with and without Coulomb repulsion

$$R_{1,0}(\rho) = \frac{\xi}{\sqrt{\pi}} \rho e^{-\frac{1}{2}\xi\rho^2}; \qquad 2\pi \int_0^\infty \rho |R(\rho)|^2 d\rho = 1,$$

where for convenience the radius ρ is measured in units of the length $\lambda = \frac{\hbar^2}{2me^2}$.

A numerical solution of the lowest eigenfunction including Coulomb repulsion for $\mu = 1$ and $\xi = 10$ is shown red in Fig. 8.5, while the corresponding oscillator function is shown in blue.

The two wave functions are surprisingly close to each other. The contribution of the Coulomb potential to the energy may be approximated by first order perturbation theory as its average over the oscillator function. The result again lies surprisingly close to the exact value (in our example, up to four digits).

On the other hand, without the Coulomb potential the wave function would have been centered at any arbitrary position in the plane. The existence of the repulsive center, however, fixes its position! This suggested the construction of the so-called Laughlin-wave functions for many electrons out of oscillator wave functions.

It is worth to remark also that the lowest s-wave ($\mu = 0$) state shows in concordance with the formerly discussed theory of Kubo a motion concentrated on a circle (equipotential line for the Coulomb potential).

Chapter 9
Extension of the Solid-State Hamiltonian: Current–Current Interaction Terms of Order $1/c^2$

We describe here an extension of the solid-state Hamiltonian to include $1/c^2$ terms, which are responsible for current–current interactions, absent in the basic Hamiltonian of Chap. 1. We construct first a Hamiltonian for classical charged point-like particles including non-relativistic terms of order $1/c^2$ using the scheme of Landau–Lifschitz in order to avoid the diverging self-interactions in the Coulomb gauge. The second quantization of this Hamiltonian contains besides the usual "charge $\leftrightarrow$ charge" interaction a similar "transverse current $\leftrightarrow$ transverse current" interaction. This Hamiltonian might serve for further developments of solid-state theory.

9.1 Classical Approach

We began this compendium with the definition of the solid state as being described by a quantum mechanical Hamiltonian of electrons and ions interacting through Coulomb forces. This traditional textbook Hamiltonian is not at all self-evident. It stems not from a classical Hamiltonian obtained from a given Lagrangian and a quantization according to the "Poisson brackets to commutators" recipe, as it was the case for a single particle in the presence of a potential. A consistent Lagrangian formalism for classical charged point-like particles interacting with the electromagnetic field, whose sources are these themselves is actually missing. One has to exclude by hand the self-interaction from the Lorentz force in order to obtain some meaningful equations (see Sect. 11.6). Nevertheless, it was a useful construction, which in its second quantized version may be justified as neglecting any $1/c$ correction in the non-relativistic quantum electrodynamics (QED) restricted to states without photons.

L. A. Bányai, *A Compendium of Solid State Theory*,
https://doi.org/10.1007/978-3-030-37359-7_9

Actually, one may go also a step further by including also the terms of order $1/c^2$ we want to describe here, although it was not yet used in solid-state theory. One must stress, however, that what we do in this chapter is just a construction and not a proof. It will be checked in the following chapter as a proper approximation of the non-relativistic QED.

One starts the construction with the classical Lagrangian of a single electron in an external field produced by some external sources ρ^{ext} and $\mathbf{i}^{ext}$

$$L(\mathbf{r},\dot{\mathbf{r}}) = \frac{m\dot{\mathbf{r}}^2}{2} - e\phi^{ext}(\mathbf{r},t) + \frac{e}{c}\mathbf{A}^{ext}(\mathbf{r},t)\dot{\mathbf{r}}.$$

In the Coulomb gauge

$$\nabla\mathbf{A}^{ext}(\mathbf{r},t) = 0$$

the potentials are

$$\phi(\mathbf{r},t)^{ext} = \int d\mathbf{x}\frac{\rho^{ext}(\mathbf{x},t)}{|\mathbf{r}-\mathbf{x}|}; \qquad \mathbf{A}^{ext}(\mathbf{r},t) = \int d\mathbf{x}\frac{\mathbf{i}_\perp^{ext}(\mathbf{x},t-|\mathbf{r}-\mathbf{x}|/c)}{c|\mathbf{r}-\mathbf{x}|},$$

where $\rho^{ext}(\mathbf{x},t)$ is the external charge density, while $\mathbf{i}_\perp^{ext}(\mathbf{x},t)$ is the external transverse ($\nabla\mathbf{i}_\perp^{ext} = 0$) current density

$$\mathbf{i}_\perp^{ext}(\mathbf{x},t) \equiv \mathbf{i}^{ext}(\mathbf{x},t) + \frac{1}{4\pi}\nabla\int d\mathbf{x}'\frac{\nabla'\mathbf{i}^{ext}(\mathbf{x}',t)}{|\mathbf{x}-\mathbf{x}'|}.$$

Next, we retain the lowest approximation in $1/c$ of the retarded current density

$$\mathbf{A}^{ext}(\mathbf{r},t) \approx \int d\mathbf{x}\frac{\mathbf{i}_\perp^{ext}(\mathbf{x},t)}{c|\mathbf{r}-\mathbf{x}|},$$

which gives rise to an $1/c^2$ term in the Lagrangian.

If the source of the fields is a single point particle of charge e' at $\mathbf{x}(t)$ having the velocity $\dot{\mathbf{x}}(t)$, then

$$\rho^{ext}(\mathbf{x},t) = e'\delta(\mathbf{x}-\mathbf{x}(t)), \qquad \mathbf{i}^{ext}(\mathbf{x},t) = e'\dot{\mathbf{x}}(t)\delta(\mathbf{x}-\mathbf{x}(t)),$$

with

$$\phi^{ext}(\mathbf{r},t) = \frac{e'}{|\mathbf{r}-\mathbf{x}(t)|}$$

and

$$\mathbf{A}^{ext}(\mathbf{r},t) = \frac{e'}{c}\left[\frac{\dot{\mathbf{x}}(t)}{|\mathbf{r}-\mathbf{x}(t)|} - \frac{1}{4\pi}\int d\mathbf{x}\frac{1}{|\mathbf{r}-\mathbf{x}|}\nabla\left(\dot{\mathbf{x}}(t)\nabla\frac{1}{|\mathbf{x}-\mathbf{x}(t)|}\right)\right] .$$

Therefore the Lagrangian of the electron in the field of the another electron, in this approximation, is

$$\mathscr{L}(\mathbf{r},\dot{\mathbf{r}};\mathbf{x},\dot{\mathbf{x}}) = \frac{m_i\dot{\mathbf{r}}^2}{2} - \frac{ee'}{|\mathbf{r}-\mathbf{x}(t)|}$$
$$+\frac{ee'\dot{\mathbf{r}}}{c^2}\left[\frac{\dot{\mathbf{x}}(t)}{|\mathbf{r}-\mathbf{x}(t)|} - \frac{1}{4\pi}\int d\mathbf{x}\frac{1}{|\mathbf{r}-\mathbf{x}|}\nabla\left(\dot{\mathbf{x}}(t)\nabla\frac{1}{|\mathbf{x}-\mathbf{x}(t)|}\right)\right].$$

By generalization one obtains for a system of N charged particles the total Lagrange function

$$L=\sum_i\frac{m_i}{2}\mathbf{v_i}^2 - \sum_{i>j}\frac{e_ie_j}{|\mathbf{r_i}-\mathbf{r}_j|}$$
$$+\sum_{i>j}\frac{e_ie_j}{c^2}\mathbf{v}_i\left[\frac{\mathbf{v}_j}{|\mathbf{r}_i-\mathbf{r}_j|} + \frac{1}{4\pi}\int d\mathbf{x}\frac{1}{|\mathbf{r}_i-\mathbf{x}|}\nabla\left(\mathbf{v}_j\nabla\frac{1}{|\mathbf{x}-\mathbf{r}_j|}\right)\right].$$

Since this Lagrangian contains velocity dependent terms besides the kinetic energy, the standard canonical formalism strictly speaking cannot be applied, because it will produce relations between the canonical momenta. Nevertheless, sticking to the lowest order in $1/c$ we have

$$\mathbf{p}_i = \frac{\delta L}{\delta\dot{\mathbf{r}}_i} \approx m\dot{\mathbf{r}}_i$$

and therefore we may still remain in the frame of the standard canonical formalism. The resulting classical Hamiltonian is

$$H = \sum_i\frac{\mathbf{p_i}^2}{2m_i} + \sum_{i>j}\frac{e_ie_j}{|\mathbf{r_i}-\mathbf{r}_j|} \tag{9.1}$$
$$-\sum_{i>j}\frac{e_ie_j}{c^2m_im_j}\mathbf{p}_i\left[\frac{\mathbf{p}_j}{|\mathbf{r}_i-\mathbf{r}_j|} - \frac{1}{4\pi}\int d\mathbf{x}\frac{1}{|\mathbf{r}_i-\mathbf{x}|}\nabla\left(\mathbf{p}_j\nabla\frac{1}{|\mathbf{x}-\mathbf{r}_j|}\right)\right]$$

including $1/c^2$ terms.

The next step is to quantize this Hamiltonian. For the sake of simplicity we consider from here on a single sort of particles (electrons) of mass m and charge e.

9.2 The Second Quantized Version

By introducing the charge and current densities:

$$\rho(\mathbf{x}) = \sum_i e\delta(\mathbf{x} - \mathbf{r}_i); \qquad \mathbf{i}(\mathbf{x}) = \sum_i \frac{e}{m}\mathbf{p}_i\delta(\mathbf{x} - \mathbf{r}_i)$$

one would be tempted to rewrite the classical Hamiltonian Eq. 9.1 as

$$\sum_i \frac{1}{2m}\mathbf{p_i}^2 + \frac{1}{2}\int d\mathbf{x}\int d\mathbf{x}'\frac{\rho(\mathbf{x})\rho(\mathbf{x}')}{|\mathbf{x} - \mathbf{x}')|} - \frac{1}{2}\int d\mathbf{x}\int d\mathbf{x}'\frac{\mathbf{i}_\perp(\mathbf{x})\mathbf{i}_\perp(\mathbf{x}')}{c^2|\mathbf{x} - \mathbf{x}'|}, \tag{9.2}$$

where $\mathbf{i}_\perp(\mathbf{x})$ is the transverse part of the current density

$$\mathbf{i}_\perp(\mathbf{r}, t) \equiv \mathbf{i}(\mathbf{r}, t) + \frac{1}{4\pi}\nabla\int d\mathbf{r}'\frac{\nabla'\mathbf{i}(\mathbf{r}', t)}{|\mathbf{r} - \mathbf{r}'|}.$$

However, due to the divergent self-interaction of point-like classical particles this expression is not meaningful, even without the $1/c^2$ terms.

The quantum mechanical version of this Hamiltonian for a system of identical particles (here fermions) the problem is, however, milder. One cannot identify individual particles and therefore the self-interaction is at least not obvious and one may eliminate it partially in the second quantization formalism by considering a "normal ordering" of the operators in the Hamiltonian, as it was done also in the case of the Coulomb interaction. This ordering of the creation operators to the left side and annihilation operators to the right side eliminates the interaction in states that contain less than two particles. Therefore we may proceed with the second quantization formulation of the theory directly form the last symbolic expression Eq. 9.2. One has to introduce the second quantized charge and the transverse part of the current density operators expressed in terms of second quantized wave functions $\psi(\mathbf{x})$ for fermions with spin 1/2, but for simplicity of the notations we omit the spin index

$$\rho(\mathbf{x}) = e\psi^+(\mathbf{x})\psi(\mathbf{x})$$

$$\mathbf{i}(\mathbf{x}) = \frac{e}{2m}\psi^+(\mathbf{x})\frac{\hbar}{\imath}\nabla\psi(\mathbf{x}) + h.c.$$

and "normal order" the second quantized electron wave functions in the interaction part of the Hamiltonian. The resulting quantum mechanical Hamiltonian $\mathbf{H}$ in the second quantized formalism looks then as

$$\mathbf{H} = -\int d\mathbf{x}\psi^+(\mathbf{x})\frac{\hbar^2}{2m}\nabla^2\psi(\mathbf{x}) + \frac{1}{2}\int d\mathbf{x}\int d\mathbf{x}'\frac{N\left[\rho(\mathbf{x})\rho(\mathbf{x}')\right]}{|\mathbf{x}-\mathbf{x}'|} - \frac{1}{2}\int d\mathbf{x}\int d\mathbf{x}'\frac{N\left[\mathbf{i}_\perp(\mathbf{x})\mathbf{i}_\perp(\mathbf{x}')\right]}{c^2|\mathbf{x}-\mathbf{x}'|} \tag{9.3}$$

where the symbol $N[..]$ means the normal ordering of the operators. Explicitly, the Coulomb term looks, as usual

$$\frac{1}{2}\int d\mathbf{x}\int d\mathbf{x}'\psi^+(\mathbf{x})\psi^+(\mathbf{x}')\frac{e^2}{|\mathbf{x}-\mathbf{x}'|}\psi(\mathbf{x}')\psi(\mathbf{x}) \ , \tag{9.4}$$

while, due to the additional integrals in the definition of the transverse part, the current–current term has a simple expression only in the discrete $\mathbf{k}$-space basis (plane waves with periodical boundary conditions in a cube of volume Ω). It looks explicitly as

$$-\frac{e^2\hbar^2}{m^2c^2\Omega}\sum_{\sigma,\sigma'=\pm 1}\sum_{\mathbf{k},\mathbf{p},\mathbf{q}}\frac{2\pi}{q^2}\left(\mathbf{kp}-\mathbf{qk}\frac{1}{q^2}\mathbf{qp}\right)a^+_{\mathbf{k},\sigma}a^+_{\mathbf{p},\sigma'}a_{\mathbf{p}+\mathbf{q},\sigma'}a_{\mathbf{k}-\mathbf{q},\sigma} \, . \tag{9.5}$$

Equation 9.3 defines the second quantized Hamiltonian of electromagnetic interacting electrons of order $1/c^2$. It includes a (transverse) current–current interaction.

We still have to add the interaction with classical external fields. These fields, sometimes called as the applied ones, have their sources far away from the system in consideration and are the basic ingredients of any electromagnetic experiment. One may add an external potential $V_{ext}(\mathbf{x})$ to the potential energy of each charged particle, whereas a classical external vector potential $\mathbf{A}^{ext}(\mathbf{x}, t)$, according to the minimal rule implies the replacement of $-\imath\hbar\nabla$ by $-\imath\hbar\nabla - \frac{e}{c}\mathbf{A}^{ext}$. Its motivation lies in the fact that in the presence of an external magnetic field the velocity of a classical charged particle is related to the canonical momentum by $m\dot{\mathbf{x}} = \mathbf{p} + \frac{e}{c}\mathbf{A}_{ext}(\mathbf{x}, t)$. This implies not only the modification of the kinetic energy term, but also the modification of the current density $\mathbf{i}_\perp(\mathbf{x})$ in the current–current term of the Hamiltonian.

Therefore, in the presence of external fields our Hamiltonian looks as

$$\mathbf{H} = -\int d\mathbf{x}\psi^+(\mathbf{x})\left[\left(\frac{\hbar}{\imath}\nabla - \frac{e}{c}\mathbf{A}_{ext}(\mathbf{x}, t)\right)^2 + V_{ext}(\mathbf{x}, t)\right]\psi(\mathbf{x}) + \frac{1}{2}\int d\mathbf{x}\int d\mathbf{x}'\frac{N\left[\rho(\mathbf{x})\rho(\mathbf{x}')\right]}{|\mathbf{x}-\mathbf{x}')|} - \frac{1}{2}\int d\mathbf{x}\int d\mathbf{x}'\frac{N\left[\mathbf{j}_\perp(\mathbf{x})\mathbf{j}_\perp(\mathbf{x}')\right]}{c^2|\mathbf{x}-\mathbf{x}'|} . \tag{9.6}$$

The current density operator $\mathbf{j}(\mathbf{x}, t)$, whose average is of interest, is modified accordingly

$$\mathbf{j}(\mathbf{x}, t) = \frac{e}{2m}\left(\psi^{+}(\mathbf{x})\left(\frac{\hbar}{\imath}\nabla - \frac{e}{c}\mathbf{A}_{ext}(\mathbf{x}, t)\right)\psi(\mathbf{x}, t) + h.c.\right). \tag{9.7}$$

In the Hartree approximation, taking into account also the presence of external potentials the above Hamiltonian looks as

$$\mathbf{H}_{Hartree} = \int d\mathbf{x}\psi^{+}(\mathbf{x}))\left[\left(\frac{\hbar}{\imath}\nabla - \frac{e}{c}\mathbf{A}_{ext}(\mathbf{x}, t)\right)^{2} + eV^{ext}(\mathbf{x}, t) + eV^{int}(\mathbf{x}, t)\right]\psi(\mathbf{x})$$
$$- \frac{1}{c}\int d\mathbf{x}\mathbf{i}_{\perp}(\mathbf{x}, t)\mathbf{A}^{int}(\mathbf{x}, t) \tag{9.8}$$

with

$$V_{int}(\mathbf{x}, t) = \int d\mathbf{x}'\frac{\langle\rho(\mathbf{x}', t)\rangle}{|\mathbf{x} - \mathbf{x}'|}; \qquad \mathbf{A}_{int}(\mathbf{x}, t) = \int d\mathbf{x}'\frac{\langle\mathbf{i}_{\perp}(\mathbf{x}', t)\rangle}{|\mathbf{x} - \mathbf{x}'|}. \tag{9.9}$$

The linear terms of this Hamiltonian with respect to the external vector potential coincide with those of the Hamiltonian used in the superconductivity models of Chap. 7. Therefore the s.c. Hamiltonian Eq. 9.8 suffers of the same weaknesses.

It is important to stress that the gauge invariance is reduced in both Hamiltonians Eqs. 9.6 and 9.8 to the choice of the external fields, since the internal fields V^{int} and $\mathbf{A}^{int}$ were deduced in a given gauge—the Coulomb one. This was the situation also without the current–current term. (Think about the hydrogen atom! One cannot change the Coulomb potential acting between the electron and the proton.) Unfortunately, this circumstance is often forgotten and one does not discern between the external, internal, or the total s.c. vector potentials. As we shall see in the next chapter, gauge invariance in the classical electromagnetic theory is restricted to the equations and to the Lagrangian, but in order to construct a Hamiltonian one has to choose a definite gauge. In the operator formulation of the (relativistic or non-relativistic) QED one may even not define (operatorial!) gauge transformations at all. It was made possible only after the functional (path integral) formulation of the quantum mechanical electromagnetic theory.

It is easy to generalize the above Hamiltonian for a system consisting of electrons and ions and therefore get the quantum mechanical Hamiltonian of the solid state including terms of order $1/c^2$ including the transverse current–current interaction.

As we mentioned in the introductory Chap. 1, the justification of ignoring the velocity dependent terms in the basic solid-state Hamiltonian was the very small velocity of the electrons in comparison to the light velocity. At a first glance the current–current interaction term seems to be of the order v^2/c^2. However, generally speaking this argumentation is false. It is an everyday's experience that a very slow flow of a macroscopic number of electrons may create enormous magnetic fields. The current–current terms in the Hamiltonian are actually responsible for the diamagnetic properties of materials and are essential for the ideal diamagnetism manifested by the Meissner effect.

The Hamiltonian Eq. 9.6 allows also the formal introduction of the source of the external fields in the bulk (including the magnetic one!), as it was done in the treatment of the linear response in the presence of Coulomb interactions in the solid-state Hamiltonian. Therefore it enables also a bulk approach to the Meissner effect. (Of course, after including terms representing the BCS interaction and taking into account the spontaneous symmetry breaking.)

The classical version of the $1/c^2$ Hamiltonian Eq. 9.1 on its turn might play an important role in the classical plasma theory, where again strong but slow currents generate strong magnetic fields.

Chapter 10
Field-Theoretical Approach to the Non-Relativistic Quantum Electrodynamics

We give here the field-theoretical derivation of the Hamiltonian of the non-relativistic quantum electrodynamics in the Coulomb gauge using the Lagrange formalism. It leads to the same result as the usual derivation, where one just replaces the classical vector potential in the minimal coupling of the second quantized electron Hamiltonian by the quantized one and adds the photon energy. This derivation however illustrates the proper use of the Euler–Lagrange equations and the canonical formalism that fails if one tries to quantize the classical theory of point-like particles interacting with the electromagnetic fields. In the same time it confirms the $1/c^2$ *result of the preceding chapter on states without photons.*

As one could see, building the quantum mechanics of charged particles starting from the classical model of point-like charges in the configuration space leads after a few steps into a dead end. More than the $1/c^2$ approximation cannot be achieved. The correct approach should follow a reverse order, starting from the formulation of the non-relativistic quantum electrodynamics, followed by simplifying approximations (expansion in powers of $1/c$).

In order to construct the non-relativistic quantum mechanical Hamiltonian describing the interaction between electrons and photons, we use here a way borrowed from the relativistic quantum field theory. We proceed by three steps. The first is to build up a classical Lagrangian density out of classical fields. These include besides the electric and magnetic fields also an one-electron wave function that leads to the coupled Maxwell and Schrödinger equations with the charge and current densities defined by the Schrödinger equation The second step is to choose the Coulomb gauge in order to eliminate the spurious degrees of freedom that allows one to build a classical Hamiltonian. The last step is to quantize all the physical

L. A. Bányai, *A Compendium of Solid State Theory*,
https://doi.org/10.1007/978-3-030-37359-7_10

fields (wave function and transverse e.m. vector potential) in the Hamiltonian. This way avoids the ill-defined theory of point-like charged particles. Of course, for the sake of stability the many-body system should contain also particles of opposite charge, but the extension of the formalism to include these is trivial.

10.1 Field Theory

In the classical field theory one defines the action $\mathscr{A}$

$$\mathscr{A} = \int d\mathbf{x} \int dt \mathscr{L}(\mathbf{x}, t)$$

by a Lagrange density $\mathscr{L}(\mathbf{x}, t)$ depending on some fields $\phi_i(\mathbf{x}, t)$ and their first time and space derivatives. The variational principle $\delta A = 0$ gives rise to the generalized Euler–Lagrange equations

$$\frac{\partial}{\partial t} \frac{\delta \mathscr{L}}{\delta \dot{\phi}_i(\mathbf{x}, t)} + \frac{\partial}{\partial x_\mu} \frac{\delta \mathscr{L}}{\delta \frac{\partial \phi_i(\mathbf{x},t)}{\partial x_\mu}} - \frac{\delta \mathscr{L}}{\delta \phi_i(\mathbf{x}, t)} = 0.$$

Here the symbol ∂ means ordinary derivative, while the symbol δ means functional derivative. Two Lagrangian densities that differ by the time derivative or by the divergence of a function give rise to the same action and therefore are considered to be equivalent, taking into account the vanishing of the fields at infinity.

Actually, we have already used this formalism in Sect. 4.2 in the treatment of the classical phonon fields on the continuum.

The generalized canonical conjugate momenta for the fields $\phi_i(\mathbf{x}, t)$ are defined by

$$\Pi_{\phi_i} = \frac{\delta \mathscr{L}}{\delta \dot{\phi}_i}$$

and the Hamiltonian density is

$$\mathscr{H}(\phi, \Pi_\phi) = -\mathscr{L} + \Pi_{\phi_i} \dot{\phi}_i,$$

provided no relations (constraints) appear between the canonical conjugate momenta. Lagrangians with constraints however have to be handled with Dirac's canonical formalism that implies also a redefinition of the Poisson bracket.

10.2 Classical Maxwell Equations Coupled to a Quantum Mechanical Electron

The classical Maxwell equations are two with sources

$$\nabla \times \mathbf{B} = \frac{4\pi}{c}\mathbf{j} + \frac{1}{c}\frac{\partial}{\partial t}\mathbf{E} \tag{10.1}$$

$$\nabla \mathbf{E} = 4\pi\rho \tag{10.2}$$

and two without sources

$$\nabla \mathbf{B} = 0 \tag{10.3}$$

$$\nabla \times \mathbf{E} = -\frac{1}{c}\frac{\partial}{\partial t}\mathbf{B}. \tag{10.4}$$

The equations without sources are automatically satisfied by the introduction of the electromagnetic potentials

$$\mathbf{B} = \nabla \times \mathbf{A} \tag{10.5}$$

$$\mathbf{E} = -\nabla V - \frac{1}{c}\frac{\partial}{\partial t}\mathbf{A}. \tag{10.6}$$

Let us suppose, that the sources

$$\rho(\mathbf{x}, t) = e\psi(\mathbf{x}, t)^*\psi(\mathbf{x}, t) \tag{10.7}$$

$$\mathbf{j}(\mathbf{x}, t) = \frac{e}{2m}\psi(\mathbf{x}, t)^*\left(-\imath\hbar\nabla + \frac{e}{c}\mathbf{A}(\mathbf{x}, t)\right)\psi(\mathbf{x}, t) + c.c \tag{10.8}$$

are given by a single electron (for simplicity without spin), described by the quantum mechanical Schrödinger equation for the wave function $\psi(\mathbf{x}, t)$

$$\imath\hbar\frac{\partial}{\partial t}\psi = \left(\frac{1}{2m}\left(-\imath\hbar\nabla + \frac{e}{c}\mathbf{A}(x, t)\right)^2 + eV(x, t)\right)\psi. \tag{10.9}$$

Using only the Schrödinger equation the sources satisfy the continuity equation (required also by consistency)

$$\nabla\mathbf{j} + \frac{\partial}{\partial t}\rho = 0. \tag{10.10}$$

The electromagnetic potentials and the wave function are not uniquely defined, they allow a simultaneous gauge transformation

$$V(\mathbf{x},t) \to V(\mathbf{x},t) + \frac{1}{c}\dot{\chi}(\mathbf{x},t) \tag{10.11}$$
$$\mathbf{A}(\mathbf{x},t) \to \mathbf{A}(\mathbf{x},t) - \nabla\chi(\mathbf{x},t)$$
$$\psi(\mathbf{x},t) \to \psi(\mathbf{x},t)e^{-\frac{\imath e}{\hbar c}\chi(\mathbf{x},t)}$$

that do not change neither the Maxwell fields **B**, **E**, the sources ρ, $\mathbf{j}$ nor the whole system of equations Eqs. 10.1 to 10.10.

The source-less Maxwell equations 10.3 and 10.4 are satisfied automatically in terms of the electromagnetic potentials through Eqs. 10.5 and 10.6. Therefore, one should concentrate only on Eqs. 10.1, 10.2, and 10.9. Then the first problem is to find a Lagrangian giving rise to these equations in terms of the fields $V(\mathbf{x},t)$, $\mathbf{A}(\mathbf{x},t)$, and $\psi(\mathbf{x},t)$ as dynamical variables (generalized coordinates). Thereafter, one has to find the classical Hamiltonian (in the Coulomb gauge) and at the end quantize simultaneously the electron wave function $\psi(\mathbf{x})$ and the transverse vector potential $\mathbf{A}_\perp(\mathbf{x})$.

10.3 Classical Lagrange Density for the Maxwell Equations Coupled to a Quantum Mechanical Electron

In our case, the fields are $V(\mathbf{x},t)$, $\mathbf{A}(\mathbf{x},t)$, and $\psi(\mathbf{x},t)$. It is easy to see that the "photon" Lagrange density

$$\mathscr{L}_{ph}(\mathbf{x},t) = \frac{1}{8\pi}\left(\nabla V(\mathbf{x},t) + \frac{1}{c}\dot{\mathbf{A}}(\mathbf{x},t)\right)^2 - \frac{1}{8\pi}(\nabla \times \mathbf{A}(\mathbf{x},t))^2 \tag{10.12}$$

through the generalized Euler–Lagrange equations for the fields $\mathbf{A}(\mathbf{x},t)$ and $V(\mathbf{x},t)$ gives rise to the Maxwell equations 10.1, respectively, Eq. 10.2, however, without sources.

On its turn, the Lagrangian density of the "electron"

$$\mathscr{L}_e(\mathbf{x},t) = -\frac{\hbar^2}{2m}\nabla\psi^*(\mathbf{x},t)\nabla\psi(\mathbf{x},t) - \frac{\imath\hbar}{2}\left(\dot{\psi}(\mathbf{x},t)^*\psi(\mathbf{x},t) - \psi(\mathbf{x},t)^*\dot{\psi}(\mathbf{x},t)\right) \tag{10.13}$$

for the field $\psi(\mathbf{x},t)^*$ gives rise to the free Schrödinger equation (uncoupled to the e.m. fields)

$$\imath\hbar\frac{\partial}{\partial t}\psi(\mathbf{x},t) = -\frac{\hbar^2}{2m}\nabla^2\psi(\mathbf{x},t). \tag{10.14}$$

In these derivations one had to use partial integration allowed by the mentioned equivalence of Lagrangian densities.

We shall introduce the e.m. interaction between the electron and the electromagnetic field by the so-called minimal way, ensuring gauge invariance. It requires the replacements

$$\frac{\hbar}{\imath}\nabla\psi \rightarrow \left(\frac{\hbar}{\imath}\nabla + \frac{e}{c}\mathbf{A}\right)\psi \tag{10.15}$$

$$\frac{\hbar}{\imath}\frac{\partial}{\partial t}\psi \rightarrow \left(\frac{\hbar}{\imath}\frac{\partial}{\partial t} + eV\right))\psi \tag{10.16}$$

in the electron Lagrangian.

Thus, the total Lagrangian is

$$\begin{aligned}\mathscr{L} = &\frac{1}{8\pi}\left(\nabla V + \frac{1}{c}\frac{\partial}{\partial t}\mathbf{A}\right)^2 - \frac{1}{8\pi}(\nabla\times\mathbf{A})^2 \\ &- \frac{1}{2m}\left(-\frac{\hbar}{\imath}\nabla + \frac{e}{c}\mathbf{A}\right)\psi^*\left(\frac{\hbar}{\imath}\nabla + \frac{e}{c}\mathbf{A}\right)\psi \\ &- \frac{1}{2}\psi^*\left(\frac{\hbar}{\imath}\frac{\partial}{\partial t} + eV\right)\psi - \frac{1}{2}\psi\left(-\frac{\hbar}{\imath}\frac{\partial}{\partial t} + eV\right)\psi^*.\end{aligned} \tag{10.17}$$

This is obviously gauge invariant and gives rise to the correct coupled equations.

10.4 The Classical Hamiltonian in the Coulomb Gauge

Unfortunately, the above introduced Lagrangian density Eq. 10.17 is a so-called singular one. The time derivative of the variable V is not present in it and therefore the corresponding canonical momentum is vanishing, i.e., we have a constraint in the canonical formalism. Lagrangians with constraints, as we already mentioned, have to be handled with Dirac's canonical formalism that implies also a redefinition of the Poisson bracket. The simplest way out is however to use the choice of the gauge in such a way as to eliminate the spurious degrees of freedom from the Lagrangian before we could construct a Hamiltonian.

The Coulomb gauge defined by

$$\nabla\mathbf{A}(\mathbf{x}, t) = 0 \tag{10.18}$$

leaves only the physical transverse degrees of freedom of the photons and simultaneously eliminates the scalar potential in favor of the charge density

$$V(\mathbf{x}, t) = \int d\mathbf{x}'\frac{\rho(\mathbf{x}', t)}{|\mathbf{x} - \mathbf{x}'|}. \tag{10.19}$$

We shall construct the Hamiltonian the usual way, however, taking into account of the above constraint and define the canonical conjugate momenta as

$$\Pi_\psi \equiv \frac{\delta \mathscr{L}}{\delta \dot{\psi}} = \frac{\imath\hbar}{2}\psi^* \tag{10.20}$$

$$\Pi_{\psi^*} \equiv \frac{\delta \mathscr{L}}{\delta \dot{\psi}^*} = -\frac{\imath\hbar}{2}\psi \tag{10.21}$$

$$\Pi_A^\mu \equiv \frac{\mathscr{L}}{\delta \dot{A}^\mu} = \frac{1}{4\pi c}\left(\frac{\partial}{x_\mu}V + \frac{1}{c}\dot{A}\mu\right), \quad (\mu = 1,2,3) \tag{10.22}$$

$$\Pi_V \equiv 0 \tag{10.23}$$

and the Hamiltonian density is

$$\begin{aligned}\mathscr{H} &= -\mathscr{L} + \Pi_{A_\mu}\dot{A}_\mu + \Pi_\psi \dot{\psi} + \Pi_{\psi^*}\dot{\psi}^* \\ &= -\frac{1}{8\pi}\left(\nabla V + \frac{1}{c}\dot{\mathbf{A}}\right)^2 + \frac{1}{8\pi}(\nabla\times\mathbf{A})^2 + \frac{1}{4\pi c}\dot{\mathbf{A}}\left(\nabla V + \frac{1}{c}\dot{\mathbf{A}}\right) \\ &+ \frac{1}{2m}\left(-\frac{\hbar}{\imath}\nabla\psi^* + \frac{e}{c}\mathbf{A}\psi^*\right)\left(\frac{\hbar}{\imath}\nabla\psi + \frac{e}{c}\mathbf{A}\psi\right) + eV\psi^*\psi,\end{aligned} \tag{10.24}$$

or (underlining also by the notation $\mathbf{A}_\perp$ the transverse character of the vector potential)

$$\begin{aligned}\mathscr{H} &= \frac{1}{8\pi}\left(\nabla V + \frac{1}{c}\dot{\mathbf{A}}_\perp\right)^2 + \frac{1}{8\pi}(\nabla\times\mathbf{A}_\perp)^2 - \frac{1}{4\pi}\nabla V\left(\nabla V + \frac{1}{c}\dot{\mathbf{A}}_\perp\right) \\ &+ \frac{1}{2m}\left(-\frac{\hbar}{\imath}\nabla\psi^* + \frac{e}{c}\mathbf{A}_\perp\psi^*\right)\left(\frac{\hbar}{\imath}\nabla\psi + \frac{e}{c}\mathbf{A}_\perp\psi\right) + eV\psi^*\psi.\end{aligned}$$

In the Hamiltonian

$$H = \int d\mathbf{x}\mathscr{H}(\mathbf{x})$$

one may use a partial integration in order to obtain

$$\begin{aligned}H = \int d\mathbf{x}\Bigg[&\frac{1}{8\pi}\left(\nabla V + \frac{1}{c}\dot{\mathbf{A}}_\perp\right)^2 + \frac{1}{8\pi}(\nabla\times\mathbf{A}_\perp)^2 + \frac{1}{4\pi}V\nabla\left(\nabla V + \frac{1}{c}\dot{\mathbf{A}}_\perp\right) \\ &+ \frac{1}{2m}\left(-\frac{\hbar}{\imath}\nabla\psi^* + \frac{e}{c}\mathbf{A}_\perp\psi^*\right)\left(\frac{\hbar}{\imath}\nabla\psi + \frac{e}{c}\mathbf{A}_\perp\psi\right) + eV\psi^*\psi\Bigg].\end{aligned}$$

Due to the transversality of the vector potential and expressing the scalar potential through the charge density Eq. 10.19, one gets

$$H = \int d\mathbf{x} \left[\frac{1}{8\pi} \left(\nabla V + \frac{1}{c} \dot{\mathbf{A}}_\perp \right)^2 + \frac{1}{8\pi} (\nabla \times \mathbf{A}_\perp)^2 \right.$$
$$\left. + \frac{1}{2m} \left(-\frac{\hbar}{\imath} \nabla \psi^* + \frac{e}{c} \mathbf{A}_\perp \psi^* \right) \left(\frac{\hbar}{\imath} \nabla \psi + \frac{e}{c} \mathbf{A}_\perp \psi \right) \right],$$

or

$$H = \int d\mathbf{x} \left[\frac{1}{8\pi} \left(\mathbf{E}^2 + \mathbf{B}^2 \right) + \frac{1}{2m} \left(-\frac{\hbar}{\imath} \nabla \psi^* + \frac{e}{c} \mathbf{A}_\perp \psi^* \right) \left(\frac{\hbar}{\imath} \nabla \psi + \frac{e}{c} \mathbf{A}_\perp \psi \right) \right].$$

Apparently the Coulomb interaction disappeared, but actually it is contained in the energy of the longitudinal electric field. Since under the integral

$$\mathbf{E}^2 = (\nabla V)^2 + (\frac{1}{c} \dot{\mathbf{A}}_\perp)^2$$

and after a partial integration

$$(\nabla V)^2 \rightarrow -V \nabla^2 V$$

we get

$$H = \int d\mathbf{x} \left[\frac{1}{8\pi} \left(\mathbf{E}_\perp^2 + \mathbf{B}^2 \right) + \frac{1}{2} e \psi^* \psi V \right. \tag{10.25}$$
$$\left. + \frac{1}{2m} \left(-\frac{\hbar}{\imath} \nabla \psi^* + \frac{e}{c} \mathbf{A}_\perp) \psi^* \right) \left(\frac{\hbar}{\imath} \nabla \psi + \frac{e}{c} \mathbf{A}_\perp \psi \right) \right].$$

The first term represents the energy of the transverse "photon" field.

10.5 Quantization of the Hamiltonian

Starting from our classical Hamiltonian in Coulomb gauge Eq. 10.26, after the usual equal-time quantization of the anti-commuting electron wave functions

$$[\psi(\mathbf{x}, t), \psi^+(\mathbf{x}', t)]_+ = \delta(\mathbf{x} - \mathbf{x}')$$

and the introduction of the creation and annihilation operators $b^+_{\mathbf{q},\lambda}$ and $b_{\mathbf{q},\lambda}$ of photons of polarization λ and momentum $\mathbf{q}$, one defines the quantized transverse e.m. vector potential

$$\mathbf{A}_{\perp}(\mathbf{x}) = \sum_{\lambda=1,2} \sqrt{\frac{\hbar c}{\Omega}} \sum_{\mathbf{q}} \frac{1}{\sqrt{|\mathbf{q}|}} \mathbf{e}_{\mathbf{q}}^{(\lambda)} e^{-\imath \mathbf{q}\mathbf{x}} \left(b_{\mathbf{q},\lambda} + b_{-\mathbf{q},\lambda}^{+} \right) \tag{10.26}$$

taken with periodical boundary conditions. This definition brings the photon part of the Hamiltonian to a diagonal form. Here the bosonic commutators are

$$\left[b_{\mathbf{q},\lambda}, b_{\mathbf{q}',\lambda'}^{+} \right] = \delta'_{\mathbf{q},\mathbf{q}} \delta_{\lambda\lambda'}$$

and the unit vectors $\mathbf{e}_{\mathbf{q}}^{(\lambda)}$ are orthogonal to the wave vector $\mathbf{q}$ and to each other

$$\mathbf{q}\mathbf{e}_{\mathbf{q}}^{(\lambda)} = 0; \qquad \mathbf{e}_{\mathbf{q}}^{(\lambda)} \mathbf{e}_{\mathbf{q}}^{(\lambda')} = \delta_{\lambda\lambda'}; \qquad \mathbf{e}_{\mathbf{q}}^{(\lambda)} = \mathbf{e}_{-\mathbf{q}}^{(\lambda)}; \qquad (\lambda, \lambda' = 1, 2).$$

With these ingredients and the normal ordering of the operators, one gets the non-relativistic QED Hamiltonian

$$\begin{aligned} H^{QED} &= \sum_{\mathbf{q},\lambda} \hbar\omega_q b_{\mathbf{q},\lambda}^{+} b_{\mathbf{q},\lambda} \\ &+ \int d\mathbf{x} N \left[\frac{1}{2m} \left(-\frac{\hbar}{\imath} \nabla \psi^{+}(\mathbf{x}) + \frac{e}{c} \mathbf{A}_{\perp}(\mathbf{x}) \psi^{+}(\mathbf{x}) \right) \left(\frac{\hbar}{\imath} \nabla \psi(\mathbf{x}, t) + \frac{e}{c} \mathbf{A}_{\perp}(\mathbf{x}) \psi(\mathbf{x}) \right) \right] \\ &+ \frac{1}{2} \int d\mathbf{x} \int d\mathbf{x}' \psi^{+}(\mathbf{x}) \psi^{+}(\mathbf{x}') \frac{e^2}{|\mathbf{x} - \mathbf{x}'|} \psi(\mathbf{x}') \psi(\mathbf{x}), \end{aligned} \tag{10.27}$$

where according to the general recipe of second quantization a normal ordering $N(\ldots)$ had to be introduced also with respect to the photon creation and annihilation operators $b_{\mathbf{q},\lambda}^{+}$, $b_{\mathbf{q}}$ in the Hamiltonian, and the photon frequency is $\omega_q = c|q|$.

This non-relativistic QED Hamiltonian coincides with the standard one obtained directly from the second quantized Hamilton operator of electrons interacting with a classical electromagnetic field in the Coulomb gauge, after the quantization of the transverse vector potential and adding the energy of the photons.

Our derivation has shown, how a field-theoretical treatment allows the use of the Lagrange formalism in deriving a non-relativistic quantum mechanical many-body theory of charged particles interacting with photons avoiding the problems linked to point-like classical charges.

10.6 Derivation of the $1/c^2$ Hamiltonian

In this framework one can show that the tedious construction of Chap. 9 is indeed equivalent to the $1/c^2$ approximation of the non-relativistic QED on states without photons. Then, if we want to retain only contributions up to order $1/c^2$, we may omit from the beginning the "seagull" term $\frac{e^2}{c^2} \int \psi^{+}\psi AA$. Being itself of order $1/c^2$, it

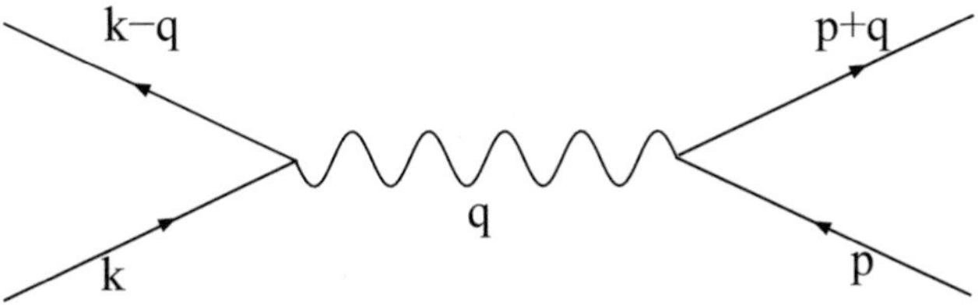

Fig. 10.1 The basic current–current (transverse photon exchange) graph in QED

may have only even higher order non-vanishing matrix elements in this subspace and besides the standard Coulomb term one is left only with the photon-current interaction $-\frac{1}{c}\int i_\perp A$.

In the S-matrix theory of adiabatic perturbations within this subspace, any Feynman diagram may be constructed using only the Coulomb vertex Fig. 3.3 and the basic graph having four electron legs and two current-photon vertices connected by a photon propagator as shown in Fig. 10.1. This graph differs from that of Fig. 3.3 by two features: the wavy line corresponds to the transverse photon propagator in the 4-dimensional Fourier space ω, $\mathbf{q}$

$$\frac{1}{q^2 - \omega^2/c^2 - \imath 0}(\delta_{\mu,\nu} - \frac{q_\mu q_\nu}{q^2}); \qquad (\mu, \nu = 1, 2, 3)$$

and the vertices contain momentum factors $-\imath\frac{e\hbar}{m}p_\mu$, $\quad -\imath\frac{e\hbar}{m}p_\nu$ due to the replacement of the charge densities by the currents. After neglecting the term $-\omega^2/c^2$ in the denominator of the photon propagator (i.e., ignoring retardation and implicitly eliminating corrections of higher order as $1/c^2$ already contained in the vertex parts), the photon propagator looks as

$$\frac{1}{q^2}(\delta_{\mu,\nu} - \frac{q_\mu q_\nu}{q^2}); \qquad (\mu, \nu = 1, 2, 3).$$

Since no pole survived, the $-\imath 0$ term could have been also ignored. Then one may convince oneself that this graph coincides with the basic vertex of the S-matrix of the $1/c^2$ theory including the transverse current–current interaction of the previous section.

Strange enough, although the non-relativistic QED and its correct $1/c^2$ approximation were well-known already in the sixties of the last century, they were used only in the treatment of the optical phenomena and completely ignored in the treatment of magnetism.

Chapter 11
Shortcut of Theoretical Physics

Here we offer a list of definitions and formulas that may help the reader to refresh his knowledge of theoretical physics.

11.1 Classical Mechanics

Lagrange function of a point-like particle in the presence of a potential $U(\mathbf{x}, t)$:

$$L(\mathbf{x}, \dot{\mathbf{x}}) = \frac{m}{2}\dot{\mathbf{x}}^2 - U(\mathbf{x}, t).$$

Euler equations:

$$\frac{\partial}{\partial t}\frac{\partial L}{\partial \dot{\mathbf{x}}} - \frac{\partial L}{\partial \mathbf{x}} = 0.$$

Hamilton function:

$$\mathbf{q} = \mathbf{x}; \qquad \mathbf{p} = \frac{\partial L}{\partial \dot{\mathbf{x}}}$$

$$H(\mathbf{p}, \mathbf{q}) = -L + \mathbf{p}\dot{\mathbf{q}}.$$

Poisson brackets:

$$\{f, g\} = \frac{\partial f}{\partial \mathbf{q}}\frac{\partial g}{\partial \mathbf{p}} - \frac{\partial g}{\partial \mathbf{q}}\frac{\partial f}{\partial \mathbf{p}}.$$

L. A. Bányai, *A Compendium of Solid State Theory*,
https://doi.org/10.1007/978-3-030-37359-7_11

Equations of motion:

$$\dot{\mathbf{q}} = \{H, q\}; \quad \dot{\mathbf{p}} = \{H, p\}.$$

The Hamiltonian formulation may be extended to include the interaction with a magnetic field, leading to the velocity dependent Lorentz force! In this case the velocity is related to the canonical momentum by

$$m\dot{\mathbf{x}} = \mathbf{p} + \frac{e}{c}\mathbf{A}(\mathbf{x}, t).$$

11.2 One-Particle Quantum Mechanics

The state of a particle is defined by a wave function $\psi(\mathbf{x}, t)$ and its time evolution is given by the Schrödinger equation

$$\imath\hbar\frac{\partial}{\partial t}\psi = H(t)\psi,$$

driven by a Hamilton operator (Hamiltonian) $H(t)$. The formal solution is given by an unitary operator $U(t, 0)$

$$U(t, 0)^{+}U(t, 0) = 1,$$

as

$$\begin{aligned} \psi(t) &= U(t, 0)\psi(0) \\ \imath\hbar\frac{\partial}{\partial t}U(t, 0) &= H(t)U(t, 0) \\ U(t, 0) &= T\left\{e^{-\frac{\imath}{\hbar}\int_0^t dt' H(t')}\right\}. \end{aligned}$$

The symbolic notation $T\{\ldots\}$ means a chronological ordering of the operators in the expansion of the exponential.

If the Hamiltonian is time independent, then

$$U(t, 0) = e^{-\frac{\imath}{\hbar}Ht}.$$

Scalar product:

$$(\psi_1, \psi_2) \equiv \int d\mathbf{x}\psi_1(\mathbf{x})^*\psi_2(\mathbf{x}).$$

Matrix elements of an operator $\mathscr{A}$

$$(\psi_1, \mathscr{A}\psi_2) \equiv \int dx \psi_1(x)^* \mathscr{A}\psi_2(x) = \int dx (\mathscr{A}^+\psi_1(x))^* \psi_2(x),$$

where $\mathscr{A}^+$ is the adjoint of $\mathscr{A}$.

Average of an observable (self-adjoint operator $\mathscr{A} = \mathscr{A}^+$):

$$\langle \mathscr{A}(t) \rangle \equiv \int dx \psi(x,t)^* \mathscr{A}\psi(x,t).$$

The Heisenberg picture:

$$\mathscr{A}(t) \equiv U(t,0)^+ \mathscr{A} U(t,0)$$

and the wave function remains time independent

$$\langle \mathscr{A}(t) \rangle = \int dx \psi(x,0)^* \mathscr{A}(t)\psi(x,0)$$

or

$$\imath\hbar \frac{\partial}{\partial t}\mathscr{A}(t) = [\mathscr{A}(t), H(t)].$$

The Hamilton operator of a particle of mass m and charge e in an external classical electromagnetic field described by external (given, classical) scalar and vector potentials $V(\mathbf{x},t)$, $\mathbf{A}(\mathbf{x},t)$

$$H(t) = \frac{1}{2m}\left(-\imath\hbar\nabla + \frac{e}{c}\mathbf{A}(x,t)\right)^2 + eV(x,t).$$

Stationary problem in the presence of a time-independent potential:

$$H = -\frac{\hbar^2}{2m}\nabla^2 + eV(\mathbf{x}).$$

Eigenvalue problem: finding the eigenfunctions ϕ_i and energy eigenvalues E_i

$$H\phi_i = E_i\phi_i.$$

The system of eigenfunctions is complete and may be eigenfunction orthonormalized:

$$(\phi_i, \phi_j) = \delta_{ij}$$

$$\sum_i \phi_i(x)\phi_i(x')^* = \delta(x,x').$$

11.2.1 Dirac's "bra/ket" Formalism

States are represented symbolically as "bra" 's $\langle\Psi|$ and "ket" 's $|\Psi\rangle$.

Scalar product (*bracket*):

$$\langle\psi_1|\psi_2\rangle.$$

Eigenstates and eigenvalues:

$$H|i\rangle = E_i|i\rangle.$$

Projectors on an eigenstate:

$$|i\rangle\langle i|$$

$$\sum_i |i\rangle\langle i| = 1$$

$$H = \sum_i E_i|i\rangle\langle i|.$$

11.3 Perturbation Theory

11.3.1 Stationary Perturbation

Time-independent perturbation proportional to a small λ:

$$H = H_0 + \lambda H'.$$

Expansion in λ :

$$H_0\phi_n^{(0)} = E_n^{(0)}\phi_n^{(0)};\; H\phi_n = E_n\phi_n$$

$$\phi_n = \phi_n^{(0)} + \lambda\phi_n^{(1)} + \lambda^2\phi_n^{(2)}\ldots$$
$$E_n = E_n^{(0)} + \lambda E_n^{(1)} + \lambda^2 E_n^{(2)} + \ldots$$

$$H'_{mn} \equiv (\phi_m^{(0)}, H'\phi_n^{(0)}).$$

Without degeneracy:

$$\phi_n^{(1)} = \sum_{m(\neq n)} \frac{H'_{mn}}{E_n^{(0)} - E_m^{(0)}}$$

$$E_n^{(1)} = H'_{nn}$$

$$E_n^{(2)} = \sum_{m(\neq n)} \frac{|H'_{mn}|^2}{E_n^{(0)} - E_m^{(0)}}.$$

With degeneracy:

$$E_{n,s}^{(0)} = E_n^{(0)}; \ (s = 1, \dots S)$$

to zeroth order for the eigenstate and first order in the eigenenergy

$$\sum_{s'=1}^{S} H'_{ns,ns'} \phi_{n,s'}^{(0)} = E^{(1)} \phi_{n,s}^{(0)}.$$

11.3.2 Time-Dependent Adiabatic Perturbation

Time-dependent perturbation proportional to a small λ:

$$H(t) = H_0 + \lambda H'(t).$$

The interaction picture is defined by

$$|\psi(t)\rangle_I \equiv e^{\frac{\imath}{\hbar} H_0 t} |\psi(t)\rangle$$

and the Schrödinger equation in the interaction picture looks as

$$\imath\hbar \frac{\partial}{\partial t} |\psi(t)\rangle_I = \lambda e^{\frac{\imath}{\hbar} H_0 t} H'(t) e^{-\frac{\imath}{\hbar} H_0 t} |\psi(t)\rangle_I = \lambda H'(t)_I |\psi(t)\rangle_I.$$

The unitary evolution in the interaction picture

$$U_I(t, t_0) = T \left\{ e^{-\frac{\imath}{\hbar} \lambda \int_{t_0}^{t} H'(t')_I dt'} \right\}.$$

The S-Matrix

$$S = \lim_{t\to\infty} \lim_{t_0\to-\infty} U(t, t_0)$$

may be defined in the adiabatic perturbation theory, if one replaces $H'(t)$ by $H'(t)e^{-0|t|}$.

The asymptotic transition rate between the unperturbed eigenstates $|n\rangle$and $|m\rangle$ of H_0 is then given by

$$\mathscr{W}_{nm} \equiv \lim_{t\to\infty} \lim_{t_0\to-\infty} \frac{d}{dt} |\langle n|U(t, t_0)|m\rangle|^2.$$

The "golden rule" to second order in λ, with an adiabatic, but oscillating in time perturbation (light absorption or induced emission)

$$H'(t) = (H'e^{\imath\omega t} + h.c.)e^{-0|t|}$$

$$\mathscr{W}_{nm} = \frac{2\pi}{\hbar}\lambda^2 |H'_{nm}|^2 \delta(E_n^{(0)} - E_m^{(0)} \pm \hbar\omega).$$

11.4 Many-Body Quantum Mechanics

11.4.1 Configuration Space

Hamilton operator and wave function of N Coulomb interacting identical particles in configuration space:

$$H^{(N)} = \sum_{i=1}^{N} \left(-\frac{\hbar^2}{2m} \nabla_i^2 + U(\mathbf{x}_i) \right) + \frac{1}{2} \sum_{i\neq i'}^{N} \frac{e^2}{|\mathbf{x}_i - \mathbf{x}_{i'}|}$$

$$H^{(N)} \Phi(x_1, \cdots x_N) = E_N \Phi(x_1, \cdots x_N).$$

The wave function $\Phi(x_1, \cdots x_N)$ must be anti-symmetrical for fermions and symmetrical for bosons.

11.4.2 Fock Space (Second Quantization)

11.4.2.1 Fermions

Using Dirac's notations one constructs simple occupied states from the vacuum state:

$$|0\rangle$$

using the creation and annihilation operators $c^+, \quad c$:

$$c^+|0\rangle = |1\rangle \qquad c^+|1\rangle = 0$$
$$c|1\rangle = |0\rangle \qquad c|0\rangle = 0.$$

The occupation number operator $n \equiv c^+c$

$$c^+c|0\rangle = 0; \quad c^+c|1\rangle = |1\rangle.$$

To every one-particle state $|k\rangle$ corresponds a creation operator c_k^+. For fermions they are anti-commuting:

$$\left[c_{k'}, c_k^+\right]_+ = \delta_{k,k'}; \qquad [c_{k'}, c_k]_+ = 0.$$

Many fermion state:

$$|\Phi\rangle = c_1^+ c_2^+ \cdots c_N^+|0\rangle$$

$$c_k^+|\cdots n_k \cdots\rangle = (-1)^{\nu_k}\sqrt{1-n_k} \quad |\cdots n_k + 1 \cdots\rangle,$$
$$c_k|\cdots n_k \cdots\rangle = (-1)^{\nu_k}\sqrt{n_k} \quad |\cdots n_k - 1 \cdots\rangle$$

$$\nu_k = \sum_{i<k} n_i.$$

Second quantized wave function is constructed with the help of a complete orthonormalized system of eigenfunctions ϕ_k

$$\psi(x) = \sum_k \phi_k(x) c_k$$

$$\left[\psi(x), \psi(x')^+\right]_+ = \delta(x - x')$$
$$\left[\psi(x), \psi(x')\right]_+ = 0.$$

Out of ground state (vacuum) $|0\rangle$ one may build a many-body basis

$$|x_N, \cdots, x_1\rangle \equiv \psi(x_1)^+ \cdots \psi(x_N)^+|0\rangle$$

$$\langle x_1, \cdots, x_N|x_1', \cdots, x_N'\rangle = det\left|\delta(x_i - x_j')\right|.$$

The Hamilton operator and the eigenstate of Coulomb interacting fermions in Fock space:

$$H=\int dx\psi(x)^{+}\left[-\frac{\hbar^2}{2m}\nabla^2+U(x)\right]\psi(x)+\frac{1}{2}\int dx\int dx'\psi(x)^{+}\psi(x')^{+}\frac{e^2}{|x-x'|}\psi(x')\psi(x).$$

The relationship to the configuration space description:

$$H|\Phi_N\rangle = E_N|\Phi_N\rangle$$

$$|\Phi_N\rangle \equiv \int dx_1\cdots\int dx_N\frac{1}{\sqrt{N!}}\Phi(x_1,\cdots x_N)|x_1,\cdots,x_N\rangle.$$

Particle density:

$$n(\mathbf{x}) = \psi(\mathbf{x})^{+}\psi(\mathbf{x}).$$

Current density:

$$\mathbf{j}(\mathbf{x}) = -\frac{\imath\hbar}{2m}\psi(\mathbf{x})^{+}\nabla\psi(\mathbf{x}) + h.c.$$

Continuity equation.

$$\frac{\partial n(\mathbf{x})}{\partial t} + \nabla\mathbf{j}(\mathbf{x}) = 0.$$

11.4.2.2 Bosons

Creation and annihilation operators are defined by

$$\begin{aligned} a|0\rangle &= 0 \\ a^{+}|n\rangle &= \sqrt{n+1}|n+1\rangle \\ a|n\rangle &= \sqrt{n}|n-1\rangle \end{aligned}$$

and the occupation number operator is

$$n \equiv a^{+}a|n\rangle = n|n\rangle.$$

The creation and annihilation operators obey the commutation rules

$$\left[a_{k'}, a_k^{+}\right] = \delta_{k,k'}; \qquad [a_{k'}, a_k] = 0.$$

Many boson states:

$$\Phi = \frac{1}{\sqrt{n_1! \ldots n_N!}} (a_1^+)^{n_1} \ldots (a_N^+)^{n_N} |0\rangle \qquad \sum_i n_i = N.$$

Second quantized wave function:

$$\psi(x) = \sum_k \phi_k(x) a_k$$

$$\left[\psi(x), \psi(x')^+\right] = \delta(x - x')$$
$$\left[\psi(x), \psi(x')\right] = 0$$

$$|x_N, \cdots, x_1\rangle \equiv \psi(x_1)^+ \cdots \psi(x_N)^+ |0\rangle$$

$$\langle x_1, \ldots, x_N | x_1', \ldots, x_N' \rangle = \sum_P \delta(x_1 - x_{j_1}') \ldots \delta(x_N - x_{j_N}').$$

11.5 Density Matrix (Statistical Operator)

Pure quantum mechanical ensemble defined by the state Φ.

$$\langle \mathscr{A} \rangle = \langle \Phi | \mathscr{A} | \Phi \rangle.$$

Mixed ensemble (many possible states with associated probabilities)

$$|\Phi_\alpha\rangle \qquad with \quad probability \quad p_\alpha \qquad (0 \le p_\alpha \le 1; \quad \sum_\alpha p_\alpha = 1)$$

$$\langle \mathscr{A} \rangle = \sum_\alpha p_\alpha \langle \Phi_\alpha | \mathscr{A} | \Phi_\alpha \rangle.$$

Density matrix (Statistical operator):

$$R \equiv \sum_\alpha p_\alpha |\Phi_\alpha\rangle\langle\Phi_\alpha|$$
$$Tr\{R\} = 1 \ ; \qquad R > 0$$
$$\imath\hbar \frac{\partial R}{\partial t} = [H, R]$$
$$\langle \mathscr{A} \rangle = Tr\{\mathscr{A} R\}.$$

Micro-canonical equilibrium:

$$R_0 = \delta(H^{(N)} - E_N).$$

Grand-canonical equilibrium.

$$R_0 = \frac{e^{-\beta(H-\mu N)}}{Tr\left(e^{-\beta(H-\mu N)}\right)}.$$

Grand-canonical potential:

$$\mathscr{F}(V,T,\mu) \equiv= -k_B T \ln Z \; ; \qquad Z \equiv Tr\left\{e^{-\beta(H-\mu N)}\right\}$$

free fermions:

$$\langle a_k^+ a_k \rangle = \frac{1}{e^{\beta(e_\mathbf{k}-\mu)}+1}$$

free bosons:

$$\langle a_k^+ a_k \rangle = \frac{1}{e^{\beta(e_\mathbf{k}-\mu)}-1}.$$

11.6 Classical Point-Like Charged Particles and Electromagnetic Fields

Classical Maxwell–Lorenz equations

$$\begin{aligned}
\nabla \times \mathbf{b} &= \frac{4\pi}{c}\mathbf{j} + \frac{1}{c}\frac{\partial}{\partial t}\mathbf{e} \\
\nabla \times \mathbf{e} &= -\frac{1}{c}\frac{\partial}{\partial t}\mathbf{b} \\
\nabla \mathbf{b} &= 0 \\
\nabla \mathbf{e} &= 4\pi\rho.
\end{aligned}$$

Continuity equation:

$$\nabla \mathbf{j} + \frac{\partial}{\partial t}\rho = 0.$$

Sources (*including external ones without dynamics*):

$$\rho(\mathbf{r}, t) = \sum_i e_i \delta(\mathbf{r} - \mathbf{r}_i(t)) + \rho^{ext}(\mathbf{r}, t)$$
$$\mathbf{j}(\mathbf{r}, t) = \sum_i e_i \mathbf{v}_i(t)\delta(\mathbf{r} - \mathbf{r}_i(t)) + \mathbf{j}^{ext}(\mathbf{r}, t)$$

$$\nabla \mathbf{j}^{ext} + \frac{\partial \rho^{ext}}{\partial t} = 0.$$

Electromagnetic potentials:

$$\mathbf{b} = \nabla \times \mathbf{a}$$
$$\mathbf{e} = -\nabla v - \frac{1}{c}\frac{\partial}{\partial t}\mathbf{a}.$$

Gauge invariance:

$$v \to v + \frac{1}{c}\frac{\partial}{\partial t}\Lambda$$
$$\mathbf{a} \to \mathbf{a} - \nabla \Lambda.$$

Equations of the potentials:

$$-\nabla^2 \mathbf{a} + \nabla(\nabla \mathbf{a}) + \frac{1}{c^2}\frac{\partial^2}{\partial t^2}\mathbf{a} = \frac{4\pi}{c}\mathbf{j} - \frac{1}{c}\nabla \frac{\partial}{\partial t} v$$
$$-\nabla^2 v - \frac{1}{c^2}\nabla \frac{\partial}{\partial t}\mathbf{a} = 4\pi\rho.$$

Coulomb gauge:

$$\nabla \mathbf{a} = 0$$

$$-\nabla^2 v = 4\pi\rho$$
$$-\nabla^2 \mathbf{a} + \frac{1}{c^2}\frac{\partial^2}{\partial t^2}\mathbf{a} = \frac{4\pi}{c}\mathbf{j} - \frac{1}{c}\nabla \frac{\partial}{\partial t} v \equiv \frac{4\pi}{c}\mathbf{j}_\perp.$$

Transverse current:

$$\nabla \mathbf{j}_\perp = 0$$

$$\mathbf{j}_\perp(\mathbf{r}, t) \equiv \mathbf{j}(\mathbf{r}, t) + \frac{1}{4\pi}\nabla \int d\mathbf{r}' \frac{\nabla' \mathbf{j}(\mathbf{r}', t)}{|\mathbf{r} - \mathbf{r}'|}.$$

Solution (with $\nabla \mathbf{a}^{ext} = 0$):

$$v(\mathbf{r}, t) = \int d\mathbf{r}' \frac{\rho^{int}(\mathbf{r}', t)}{|\mathbf{r} - \mathbf{r}'|} + V^{ext}(\mathbf{r}, t)$$

$$\mathbf{a}(\mathbf{r}, t) = \int d\mathbf{r}' \frac{\mathbf{j}_{\perp}^{int}(\mathbf{r}', t - \frac{1}{c}|\mathbf{r} - \mathbf{r}'|)}{c|\mathbf{r} - \mathbf{r}'|} + \mathbf{a}^{ext}(\mathbf{r}, t).$$

Newton's equation of motion with Lorentz forces:

$$m_i \frac{d}{dt} \mathbf{v}_i = e_i \left(\mathbf{e}'(\mathbf{r}_i, t) + \frac{1}{c} \mathbf{v}_i \times \mathbf{b}'(\mathbf{r}_i, t) \right).$$

Here $\mathbf{e}'$ *and* $\mathbf{b}'$ *have to be considered as the fields without self-action! This implies that neither Lagrangian nor canonical formalism is available for point-like charges.*

For $\frac{v}{c} \to 0$, one ignores the magnetic field created by the particles themselves $\mathbf{a}^{int}$ and the particle motion may be separated:

$$H = \sum_i \left(\frac{1}{2m_i} (\mathbf{p}_i + \frac{e_i}{c} \mathbf{a}^{ext}(\mathbf{r}_i, t))^2 + e_i V^{ext}(\mathbf{r}_i, t) \right) + \frac{1}{2} \sum_{i \neq j} \frac{e_i e_j}{|\mathbf{r}_i - \mathbf{r}_j|}.$$

In this standard approximation used in solid-state theory, the magnetic field created by the charged particles is ignored!

Macroscopic fields as averages:

$$\mathbf{E} = \langle \mathbf{e} \rangle; \qquad \mathbf{B} = \langle \mathbf{b} \rangle.$$

Macroscopic Maxwell equations.

$$\nabla \times \mathbf{B} = \frac{4\pi}{c} (\langle \mathbf{j} \rangle + \mathbf{j}^{ext}) + \frac{1}{c} \frac{\partial}{\partial t} \mathbf{E}$$

$$\nabla \times \mathbf{E} = -\frac{1}{c} \frac{\partial}{\partial t} \mathbf{B}$$

$$\nabla \mathbf{B} = 0$$

$$\nabla \mathbf{E} = 4\pi (\langle \rho \rangle + \rho^{ext}).$$

The classical macroscopic theory of electromagnetism relates the average sources ($\langle \rho \rangle$, $\langle \mathbf{j} \rangle$) *to the fields* ($\mathbf{E}$, $\mathbf{B}$) *by certain phenomenological relationships and leaves the foundation of these relationships to the microscopical theories.*

The non-relativistic quantum electrodynamics (QED) was described in Chap. 10; however, in many applications elements of the relativistic QED have to be implemented like the spin and its interaction with the magnetic field (including also that created by the spins themselves).

Chapter 12
Homework

For all readers it is a useful exercise to perform the omitted details of the proofs. I recommend also some lengthy but useful homeworks an ambitious graduate student should be able to perform successfully. I would like to encourage the reader also to write his own programs and play through different funny scenarios around the examples given in the Sects. 2.6.2, 5.2 and 8.3.1.

12.1 The Kubo Formula

Start from the linear response formula of Sect. 6.1 with the perturbation caused by coupling to (time-dependent) electromagnetic potentials in the Sects. 2.6.2, 5.2, and 8.3.1

$$H'(t) = \int d\mathbf{x} \left\{ \rho(\mathbf{x}) V(\mathbf{x}, t) - \mathbf{j}(\mathbf{x}) \mathbf{A}(\mathbf{x}, t) \right\},$$

where the $\mathbf{A}^2$ term of the non-relativistic theory was ignored, since it is of second order in the field. Prove the Kubo formula for the induced average current density

$$\langle j_\mu(\mathbf{x}, t) \rangle = \sum_{\nu=1}^{3} \int_{-\infty}^{t} dt' \int_0^\beta d\lambda \int d\mathbf{x}' \langle j_\nu(\mathbf{x}', -\imath\hbar\lambda) j_\mu(\mathbf{x}, t - t') \rangle_0 E_\nu(\mathbf{x}', t')$$

using the Kubo identity

$$\left[A, e^{-\beta H} \right] = e^{-\beta H} \int_0^\beta d\lambda e^{\lambda H} \left[H, A \right] e^{-\lambda H}.$$

L. A. Bányai, *A Compendium of Solid State Theory*,
https://doi.org/10.1007/978-3-030-37359-7_12

(Check it by taking its matrix element between eigenstates of the Hamiltonian H.) Consider the case of a homogeneous field in a homogeneous system introduced adiabatically at $t = -\infty$ as e^{st} with $s \to +0$ to get the conductivity tensor

$$\sigma_{\mu\nu}(\omega) = \lim_{s\to+0}\lim_{\Omega\to\infty} \Omega \int_{-\infty}^{0} dt e^{\imath\omega t} e^{st} \int_{0}^{\beta} d\lambda \langle j_\nu(t - \imath\hbar\lambda) j_\mu(0)\rangle_0, \qquad (12.1)$$

where $j_\mu \equiv \frac{1}{\Omega}\int d\mathbf{x} j_\mu(\mathbf{x})$ and the infinite volume limit is to be understood in the thermodynamic sense, i.e., at a fixed average carrier density. (It is understood that the Kubo formula is valid only for Coulomb non-interacting electrons, i.e., the electric field is the total one.)

In the same way one may derive also the linear relation of the energy current density $\mathbf{j}^E$ to the applied electric field, under the assumption of the existence of local energy density and energy density operators $\rho^E(\mathbf{x})$, satisfying the continuity equation with the energy current density $\mathbf{j}^E(\mathbf{x})$

$$\nabla \mathbf{j}^E(\mathbf{x}) + \frac{\partial}{\partial t}\rho^E(\mathbf{x}) = 0.$$

This assumption is not at all trivial and excludes again long range Coulomb interactions.

In Coulomb interacting systems, as it was shown in Chap. 5, there is a modification of the linear response for the conductivity to take into account Coulomb interactions beyond the mean-field approach. For the last discussed case there is no such possibility.

12.2 Ideal Relaxation

Remake the above derivation by adding an ideal relaxation term

$$-\imath\hbar\frac{R - R_0}{\tau}$$

to the Liouville equation. However, now starting in equilibrium at time $t = 0$ and measuring at $t = \infty$ without any adiabaticity. You will see that you get the same formula, with the adiabatic parameter s replaced by $\frac{1}{\tau}$.

Now, consider that the unperturbed Hamiltonian H is just the one describing the motion in a constant magnetic field in the Landau gauge, i.e., compute the matrix elements with the Landau states and perform all the integrals. Do not forget that the velocity in the presence of the magnetic field is $\dot{\mathbf{x}} \equiv \frac{1}{m}(\mathbf{p} - \frac{e}{c}\mathbf{A}(\mathbf{x}))$!

You shall obtain the same expression for the conductivity tensor as we derived in Sect. 5.3 from the Boltzmann equation under an analogous ideal relaxation assumption.

12.3 Rate Equation for Bosons

Derive the transition rates for massive bosons interacting with acoustic phonons using the many-body "golden rule." Show the detailed balance relation. Within the approximation

$$\langle n_i n_j \rangle \approx \langle n_i \rangle \langle n_j \rangle$$

formulate the corresponding rate equation with discrete wave vectors **k**. Analyze its properties. Perform the thermodynamic limit and remark the problem with the particle conservation if the concentration of bosons exceeds the critical one.

12.4 Bose Condensation in a Finite Potential Well

The standard theory of Bose condensation implies the thermodynamic limit and describes an infinite homogeneous system. On the other hand, experimental evidence occurs in finite (confined) systems. One may try a quasi-classical approach to bosons in a finite (but not quantizing!) potential well to understand this aspect.

Consider the motion in the presence of a finite in depth and width, spherically symmetric, attractive ($v > 0$) potential well

$$U_0(r) = -v\left(1 - \left(\frac{r}{R}\right)^2\right)\theta(R - r)$$

($r \equiv |\mathbf{x}|$) embedded in an infinite volume.
The corresponding equilibrium distribution is

$$f(\mathbf{p}, \mathbf{x}) = \frac{1}{e^{\beta(\frac{p^2}{2m} + U(\mathbf{x}) - \mu)} - 1}$$

with the chemical potential determined by the total number of bosons

$$\int d\mathbf{x} \int \frac{d\mathbf{p}}{(2\pi\hbar)^3} f(\mathbf{p}, \mathbf{x}) = \langle N \rangle .$$

However, since one expects that at low temperatures many bosons drop in the well, one has to take into consideration that being close to each other they will interact. Usually this interaction is repulsive. Consider just a contact interaction potential $v(\mathbf{x}) = w\delta(\mathbf{x})$ $(w > 0)$ that one may treat in a self-consistent manner by the effective potential

$$U(\mathbf{x}) = U_0(\mathbf{x}) + \int d\mathbf{x}' v(\mathbf{x} - \mathbf{x}') \int \frac{d\mathbf{p}}{(2\pi\hbar)^3} f(\mathbf{p}, \mathbf{x}').$$

This potential is just a constant outside the well

$$U(\mathbf{x}) = w\langle n\rangle \qquad (for \quad r > R),$$

where $\langle n\rangle$ is the average density of bosons

$$\langle n\rangle = \int \frac{d\mathbf{p}}{(2\pi\hbar)^3} \frac{1}{e^{\beta(\frac{p^2}{2m} + w\bar{n} - \mu)} - 1}.$$

(A finite number of bosons in the well do not contribute to this equation.) A solution exists only for $\mu < wn_c$, where n_c is the critical density

$$n_c = \int \frac{d\mathbf{p}}{(2\pi\hbar)^3} \frac{1}{e^{\beta\frac{p^2}{2m}} - 1}.$$

Above this density an overall condensate should appear.

Now, it is most interesting to follow the scenario before this overall condensation occurs. For $r < R$ one gets a radius dependent density and the self-consistency equation

$$U(r) = U_0(r) + w \int \frac{d\mathbf{p}}{(2\pi\hbar)^3} \frac{1}{e^{\beta(\frac{p^2}{2m} + U(r) - \mu)} - 1}; \qquad (r < R).$$

So long $U(r) - \mu > 0$, this equation has a solution, but afterwards, obviously not. Therefore, it is reasonable to correct this equation by admitting the possibility of a local condensate density $n_0(r)$ not included in the Bose distribution

$$U(r) = U_0(r) + w\left(n_0(r) + \int \frac{d\mathbf{p}}{(2\pi\hbar)^3} \frac{1}{e^{\beta(\frac{p^2}{2m} + U(r) - \mu)} - 1}\right) \qquad (for \quad r < R).$$

The potential $U(r)$ varies monotonously and coming down from the top of the potential one might reach a radius $r_0 < R$, by which indeed

$$U(r_0) = \mu$$

and the integral over the Bose function reaches its maximal value. Due to the condition $U(r) - \mu \geq 0$ also for the points $r < r_0$, it must belong to the minimum of $U(r)$ and a condensate $n_0(r)$ must emerge

$$n_0(r) = \frac{1}{w}\,(U(r_0) - U_0(r)) \;\; (for \quad 0 < r < r_0).$$

Solve numerically the self-consistency equation for $r_0 < r < R$ (before the apparition of the condensate) to confirm this scenario. Since the transcendental self-consistency equation is local, for a numerical solution it is convenient to solve it in favor of U_0 at a given U, thus performing a simple integration over the momenta **p**. The association to a certain radius r is given then by the explicit definition of $U_0(r)$.

Correction to: A Compendium of Solid State Theory

Correction to:
L. A. Bányai, *A Compendium of Solid State Theory*, https://doi.org/10.1007/978-3-030-37359-7

The author's affiliation that is displayed in the book and on the website is not actually the author's affiliation, but rather his location (Oberursel being a city in Germany, not an institution).

"Oberursel, Germany"

Therefore, this has now been updated as:

Institut für Theoretische Physik, Goethe Universität, Frankfurt am Main, Germany

---

The updated online version of the book can be found at:
https://doi.org/10.1007/978-3-030-37359-7

L. A. Bányai, *A Compendium of Solid State Theory*,
https://doi.org/10.1007/978-3-030-37359-7_13

Printed in the United States
By Bookmasters